T0181833

# Synthese Library

Studies in Epistemology, Logic, Methodology, and Philosophy of Science

Volume 431

The aim of *Synthese Library* is to provide a forum for the best current work in the methodology and philosophy of science and in epistemology. A wide variety of different approaches have traditionally been represented in the Library, and every effort is made to maintain this variety, not for its own sake, but because we believe that there are many fruitful and illuminating approaches to the philosophy of science and related disciplines.

Special attention is paid to methodological studies which illustrate the interplay of empirical and philosophical viewpoints and to contributions to the formal (logical, set-theoretical, mathematical, information-theoretical, decision-theoretical, etc.) methodology of empirical sciences. Likewise, the applications of logical methods to epistemology as well as philosophically and methodologically relevant studies in logic are strongly encouraged. The emphasis on logic will be tempered by interest in the psychological, historical, and sociological aspects of science.

Besides monographs *Synthese Library* publishes thematically unified anthologies and edited volumes with a well-defined topical focus inside the aim and scope of the book series. The contributions in the volumes are expected to be focused and structurally organized in accordance with the central theme(s), and should be tied together by an extensive editorial introduction or set of introductions if the volume is divided into parts. An extensive bibliography and index are mandatory.

More information about this series at http://www.springer.com/series/6607

Moti Mizrahi

# The Relativity of Theory

## Key Positions and Arguments
## in the Contemporary Scientific Realism/
## Antirealism Debate

 Springer

Moti Mizrahi
College of Psychology and Liberal Arts
Florida Institute of Technology
Melbourne, FL, USA

ISSN 0166-6991          ISSN 2542-8292    (electronic)
Synthese Library
ISBN 978-3-030-58049-0       ISBN 978-3-030-58047-6    (eBook)
https://doi.org/10.1007/978-3-030-58047-6

This Springer imprint is published by the registered company Springer Nature Switzerland AG
The registered company address is: Gewerbestrasse 11, 6330 Cham, Switzerland

# Preface

The main objective of this book is to provide an introductory overview of a "hot topic" in contemporary philosophy of science, namely, the scientific realism/antirealism debate. This introductory book is meant to be accessible to students of philosophy as well as students of the sciences who might have an interest in philosophical questions about science. It is intended for anyone who has an interest in philosophical questions about the foundations of science. No background in philosophy of science is assumed or required. Rather, this book takes an argumentation approach to the scientific realism/antirealism debate in contemporary philosophy of science, which will be explained in detail in the first chapter and then employed in subsequent chapters. The emphasis on argumentation will help students to not only see what the key arguments in the debate are but also practice their argumentation and critical thinking skills. My aim is to show readers that the scientific realism/antirealism debate in contemporary philosophy of science is not only interesting but also significant. To this end, plenty of up to date examples are used to illustrate points of contention in the debate. By focusing on the scientific realism/antirealism debate in contemporary philosophy of science from an argumentation perspective, while providing current examples for illustration purposes, this book aims to fill a gap in the philosophy of science literature.

I have been thinking about realism and antirealism in philosophy of science since undertaking my graduate course of studies at the Graduate Center of the City University of New York between 2005 and 2010. Since graduating in 2010, I have been thinking and writing about the scientific realism/antirealism debate and my work on this topic has appeared in philosophy of science journals, such as *The British Journal for the Philosophy of Science, Erkenntnis, Foundations of Science, Journal for General Philosophy of Science, International Studies in the Philosophy of Science, Perspectives on Science, Studies in History and Philosophy of Science,* and *Synthese*. This book is a culmination of all these years of thinking about the key positions and arguments in this debate. In that respect, I am quite confident that there is enough original material here, either in terms of putting old arguments in new forms or in terms of putting forth new arguments for my own position, namely, Relative Realism, which will be of interest to those who are already familiar with the scientific realism/antirealism debate in contemporary philosophy of science. I take Relative Realism to be a middle ground position between scientific realism and antirealism. According to Relative Realism, we have good reasons to believe that, from a set of competing scientific theories, the more empirically successful theory is *comparatively true*, that is, closer to the truth relative to its competitors in the set.

My thinking about the scientific realism/antirealism debate has benefited from conversations in person and via email correspondence with numerous people throughout the years, for which I am grateful, including Alberto Cordero, Andrew Aberdein, Anjan Chakravartty, Ann-Sophie Barwich, Brad Wray, Catherine Wilson, Darrell Rowbottom, David Kaspar, Fabio Sterpetti, Greg Frost-Arnold, Howard Sankey, James Beebe, James McAllister, Kyle Stanford, Lydia Patton, Otávio Bueno, Samuel Schindler, Seungbae Park, Steven French, Tim Lyons, and Vasso Kindi, among others. I would also like to thank the Springer team: Lucy Fleet, Chris Wilby, and Palani Murugesan. Finally, I am grateful to an anonymous reviewer of *Synthese Library* for their helpful comments on earlier drafts of this book.

Melbourne, FL, USA                                                    Moti Mizrahi
July 2020

# Contents

# Chapter 1
# Introduction

**Abstract** The main objective of this book is to provide an introductory overview of a "hot topic" in contemporary philosophy of science, namely, the scientific realism/antirealism debate, which is accessible to students of philosophy as well as students of the sciences who might have an interest in philosophical questions about science. Unlike other books on the scientific realism/antirealism debate in contemporary philosophy of science, this book takes an argumentation approach to the debate. That is to say, rather than devote an entire book to a defense of scientific realism (of some variety or another) or antirealism (of some variety or another), I survey and evaluate what I take to be the key positions and arguments in the scientific realism/antirealism debate in contemporary philosophy of science. Also unlike many books in the contemporary scientific realism/antirealism literature, this book does not include detailed descriptions of a few case studies from the history of science. Instead, this book includes detailed analyses and evaluations of the key arguments these case studies from the history of science are supposed to motivate in the first place. The book concludes with several arguments for my own brand of scientific realism, namely, Relative Realism. I take Relative Realism to be a middle ground position between scientific realism and antirealism.

**Keywords** Argument · Canonical form (standard form) · Cogent argument · Deductive argumentation · Inductive argumentation · Invalid argument · Non-cogent argument · Sound argument · Strong argument · Unsound argument · Valid argument · Weak argument

When she appeared before the United States House Foreign Affairs Subcommittee on Europe, Eurasia, Energy, and the Environment, the climate activist, Greta Thunberg, said to U.S. lawmakers, "I don't want you to listen to me. I want you to listen to the scientists." Greta Thunberg had to urge U.S. lawmakers to "listen to the scientists" and then take action on climate change because climate change is a divisive political issue in the United States. The results of a Gallup poll from March 2019 show that 77% of Democrats believe that global warming is caused by human activities and they are concerned about it, whereas only 16% of Republicans believe that global warming is caused by human activities and they are concerned about it.

© Springer Nature Switzerland AG 2020
M. Mizrahi, *The Relativity of Theory*, Synthese Library 431,
https://doi.org/10.1007/978-3-030-58047-6_1

Most Republicans (52%) are "cool skeptics," that is, they do not believe that global warming is caused by human activities and they are not concerned about climate change.

Now, most contemporary philosophers of science would probably agree with Greta Thunberg that we should "listen to the scientists" when they talk about the causes and dangers of climate change. This is because most contemporary philosophers of science either accept scientific realism or lean toward scientific realism. According to the results of the PhilPapers Survey, which is an opinion poll that was conducted by professional philosophers to survey the opinions of academic philosophers on various philosophical topics, 75% of professional philosophers either accept or lean toward scientific realism, whereas 11% accept or lean toward scientific antirealism (Bourget and Chalmers 2014, p. 498). Among academic philosophers who are experts in General Philosophy of Science, in particular, 60% either accept or lean toward scientific realism, whereas 16% accept or lean toward scientific antirealism. This means that most philosophers of science have "a positive epistemic attitude toward the content of our best theories and models, recommending belief in both observable and unobservable aspects of the world described by the sciences" (Chakravartty 2017). Accordingly, if our best climate models show that global warming is occurring due to human activities, we have good reasons to believe that global warming is real and that human activities are in fact a contributing factor.

Despite the fact that most professional philosophers subscribe to or lean toward scientific realism, as the results of the PhilPapers Survey suggest, the scientific realism/antirealism debate rages on as influential articles and books defending either scientific realism (of some variety or another; see, for example, Dicken 2016) or scientific antirealism (of some variety or another; see, for example, Wray 2018) continue to be published regularly. Unlike other books on the scientific realism/antirealism debate in contemporary philosophy of science, however, I would like to take a somewhat different approach to this debate in this book. Instead of devoting an entire book to a defense of scientific realism (of some variety or another) or antirealism (of some variety or another), I survey and evaluate what I take to be the key positions and arguments in the scientific realism/antirealism debate in contemporary philosophy of science. Also unlike many books in the contemporary scientific realism/antirealism literature, this book does not include detailed descriptions of a few case studies from the history of science.[1] Instead, this book includes detailed analyses and evaluations of the key arguments these case studies from the history of science are supposed to motivate in the first place.

---

[1] In Mizrahi (2018) and Mizrahi (2020), I discuss the problems with the method of using case studies as evidence for philosophical theses about science in much more detail.

## 1.1 An Argumentation Approach to the Scientific Realism/ Antirealism Debate

The main objective of this book is to provide an introductory overview of a "hot topic" in contemporary philosophy of science, namely, the scientific realism/antirealism debate, which is accessible to students of philosophy as well as students of the sciences who might have an interest in philosophical questions about science. This book is modeled after recent books in philosophy that aim to introduce students to philosophical problems, questions, and debates through argumentation. A few recent examples of books that take an argumentation approach to philosophy include the following:

- *Just the Arguments: 100 of the Most Important Arguments in Western Philosophy*, edited by Michael Bruce and Steven Barbone (Wiley-Blackwell, 2011).
- *What is the Argument? An Introduction to Philosophical Argument and Analysis* by Maralee Harrell (The MIT Press, 2016).
- *For the Sake of Argument: How to Do Philosophy* by Robert M. Martin (Broadview Press, 2017).
- *Bad Arguments: 100 of the Most Important Fallacies in Western Philosophy*, edited by Robert Arp, Steven Barbone, and Michael Bruce (Wiley-Blackwell, 2018).

Like these four books, this book attempts to cut through dense philosophical prose and present just the arguments for and against scientific realism as well as other key positions in the scientific realism/antirealism debate in contemporary philosophy of science. What I take to be key arguments in the contemporary scientific realism/ antirealism debate are presented in *canonical form*, which is also known as *standard form*, that is, in numbered premises followed by a conclusion, with objections to each argument as well as key quotations that provide references to seminal works in philosophy of science. Accordingly, by *"argument"* is meant a connected series of statements in which some statements (that is, at least one statement) are supposed to provide evidence for, or reasons to accept, another statement. That last statement is called a *conclusion*, whereas the statements that are supposed to provide evidence for the conclusion, or reasons to accept it, are called *premises*.

To illustrate this argumentation approach, here is an example of an argument in canonical or standard form (that is, in numbered premises followed by a conclusion):

(P1) All academic philosophers are scientific realists.
(P2) Nora Berenstain is an academic philosopher.

Therefore,

(C) Nora Berenstain is a scientific realist.

The statements labeled as (P1) and (P2) are the *premises* of this argument and the statement labeled as (C) is the *conclusion* of this argument. The premises of an argument are supposed to provide supporting evidence for the conclusion. To put it another way, the premises of an argument are supposed to give reasons to believe that the conclusion is true. In this case, (P1) and (P2) are supposed to provide supporting evidence for (C). In other words, (C) is supposed to follow logically from (P1) and (P2). This argument is an instance of *deductive argumentation*. In a deductive argument, the premises purport to provide logically conclusive support for the conclusion. In the aforementioned argument, the premises, namely, (P1) and (P2), do in fact support the conclusion conclusively, such that (C) must be true if (P1) and (P2) are true. A deductive argument in which the premises successfully provide logically conclusive support for the conclusion is said to be a *valid* argument. If the premises of a deductive argument purport to provide logically conclusive support for the conclusion, but fail to do so, the argument is said to be an *invalid* argument.

From the results of the PhilPapers Survey, we have some empirical evidence suggesting that only 75% of academic philosophers either accept or lean toward scientific realism, not 100% of academic philosophers (Bourget and Chalmers 2014, p. 498), as (P1) states. Since we have some evidence suggesting that (P1) is false, the aforementioned deductive argument is valid, but it cannot be said to be *sound*. For a deductive argument to be sound, it must be valid as well as have all true premises. A valid argument with even one false premise is said to be *unsound*. Given the results of the PhilPapers Survey, the above argument cannot be said to be sound because, although valid, it has a false premise, namely, (P1).

In addition to deductive arguments, some of the key arguments in the scientific realism/antirealism debate in contemporary philosophy of science are supposed to be non-deductive or inductive arguments. In non-deductive or *inductive argumentation*, the premises purport to provide probable, rather than logically conclusive, support for the conclusion. Here is an example of an inductive argument in canonical (or standard) form:

(P1) 75% of surveyed academic philosophers are scientific realists.
(P2) Nora Berenstain is an academic philosopher (who was not surveyed).

Therefore,

(C) Nora Berenstain is a scientific realist.

In this case, (P1) and (P2) are supposed to provide some supporting evidence for (C), but not logically conclusive evidence as in the previous example of a deductive argument. To put it another way, (P1) and (P2) are supposed to provide strong reasons, but not conclusive reasons, to believe that (C) is true. This kind of inductive argumentation is known as a *statistical syllogism* or *inductive prediction*. In inductive argumentation, the premises purport to make the truth of the conclusion more likely or probable, but not absolutely guaranteed. In this case, if it is true that 75% of surveyed academic philosophers are scientific realists, and if it is also true that Nora Berenstain is an academic philosopher, then it is likely also true that Nora Berenstain is a scientific realist. In other words, if the premises (P1) and (P2) are true, (C) is probably true. An inductive argument in which the premises successfully provide probable support for the conclusion is said to be a *strong* argument. If the premises of an inductive argument purport to provide probable support for the conclusion, but fail to do so, the argument is said to be a *weak* argument. In inductive prediction, for instance, given that $X$ percent of sampled things, $F$s, have a particular property, $G$, we are entitled to conclude that, with a probability of $X$ percent, a new $F$ that has not been observed or surveyed yet will also have the property, $G$, provided that $X$ is greater than 50%, and there is no evidence that the new $F$ is unlike previously observed $F$s (Schurz 2019, p. 2). Accordingly, given that 75% of surveyed academic philosophers are scientific realists, we are justified in concluding, with a probability of 75%, that Nora Berenstain is a scientific realist, given that she is an academic philosopher, even though she was not surveyed.

From the results of the PhilPapers Survey, we have some empirical evidence suggesting that 75% of academic philosophers either accept or lean toward scientific realism (Bourget and Chalmers 2014, p. 498), and so we have evidence suggesting that (P1) is true. Assuming that (P2) is true as well, that is, assuming that there really is an academic philosopher out there whose name is Nora Berenstain, the aforementioned inductive argument can be said to be a *cogent* argument.[2] A cogent argument is a strong argument with all true premises. A strong argument with even one false premise is said to be a *non-cogent* argument. Given the results from the PhilPapers Survey, the above argument can be said to be a cogent argument because it is strong and (P1) and (P2) are in fact true.

Throughout this book, key arguments in the scientific realism/antirealism debate in contemporary philosophy of science are presented in this format (that is, in canonical or standard form) in order to make their logical form and their premises clear to readers. For, in general, any argument can fail in two ways. First, an argument fails when the conclusion does not follow from the premises even if the premises of the argument are true. For example, suppose that what we know about Nora Berenstain is not that she is an academic philosopher, but rather that she is a scientific realist. In that case, the conclusion that Nora Berenstain is an academic philosopher would not follow from our premises. That is to say, the following deductive argument is an invalid argument:

(P1) All academic philosophers are scientific realists.
(P2) Nora Berenstain is a scientific realist.

Therefore,

(C) Nora Berenstain is an academic philosopher.

This deductive argument is invalid because, even if the premises of this argument were true, the conclusion would not necessarily follow from those premises. For premise (P1) tells us that all academic philosophers are scientific realists, but it does not tell us that all scientific realists are academic philosophers. So, even if Nora Berenstain is a scientific realist, it does not necessarily follow that she is an

---

[2] In a co-authored paper, Nora Berenstain and James Ladyman (2012) defend a version of scientific realism known as Ontic Structural Realism (OSR). On Structural Realism, see Chap. 3 (Sect. 3.5).

academic philosopher because (P1) tells us that all academic philosophers are scientific realists, but not that all scientific realists are academic philosophers.

Inductive arguments can also fail in this way. For example, suppose that the results of our survey show that 25% of academic philosophers are scientific realists. In that case, the conclusion that Nora Berenstain, who is an academic philosopher who was not surveyed, is a scientific realist would not follow from our premises. That is to say, the following inductive argument is a weak argument:

(P1) 25% of surveyed academic philosophers are scientific realists.
(P2) Nora Berenstain is an academic philosopher (who was not surveyed).

Therefore,

(C) Nora Berenstain is a scientific realist.

This inductive argument is a weak argument because, even if the premises of this argument were true, they would not make the conclusion more probable or likely to be true. Given the premises, the probability that Nora Berenstain (or any other academic philosopher chosen at random, for that matter) is a scientific realist is rather low, that is, merely 25%.

Second, an argument can fail when one or more of its premises is not true. In that respect, a deductive argument could be valid insofar as the premises, if true, would provide conclusive support for the conclusion, but one of those premises is in fact false. For example, the following deductive argument is an unsound argument:

(P1) All academic philosophers are scientific realists.
(P2) Nora Berenstain is an academic philosopher.

Therefore,

(C) Nora Berenstain is a scientific realist.

The premises of this deductive argument, if true, would provide conclusive support for the conclusion, so the argument is valid. However, since (P1) is false, as the results of the PhilPapers Survey (Bourget and Chalmers 2014, p. 498) suggest, the argument cannot be said to be a sound argument.

Likewise, an inductive argument could be strong insofar as the premises, if true, would make the conclusion more probable to likely to be true, but one of those premises is in fact false. For example, the following inductive argument is a non-cogent argument:

(P1) 95% of surveyed academic philosophers are scientific realists.
(P2) Nora Berenstain is an academic philosopher (who was not surveyed).

Therefore,

(C) Nora Berenstain is a scientific realist.

The premises of this inductive argument, if true, would make the truth of the conclusion more likely or probable, so the argument is strong. However, since (P1) is false, as the results of the PhilPapers Survey (Bourget and Chalmers 2014, p. 498) suggest, the argument cannot be said to be a cogent argument.

To sum up, we can broadly distinguish between two kinds of arguments: deductive arguments and non-deductive (or inductive) arguments. In deductive argumentation, a premise (or premises) purports to provide logically conclusive support for a conclusion. Deductive arguments in which the premises succeed in providing logically conclusive support for the conclusion are said to be *valid* arguments; otherwise, they are said to be *invalid* arguments. A deductive argument that is valid and all of its premises are true is said to be a *sound* argument; if even a single premise is false, it is said to be an *unsound* argument. In non-deductive or inductive argumentation, a premise (or premises) purports to provide probable support for a conclusion. Inductive arguments in which the premises succeed in providing probable support for the conclusion are said to be *strong* arguments; otherwise, they are said to be *weak* arguments. An inductive argument that is strong and all of its premises are true is said to be a *cogent* argument; if even a single premise is false, it is said to be a *non-cogent* argument.

It would be useful to have a generic decision procedure to follow when we have an argument to analyze and evaluate. The following is a generic decision procedure for the analysis and evaluation of arguments in natural language.

- **Step 1**: Identify the *conclusion* of the argument and then the evidence or reasons that are supposed to support that conclusion. The evidence or reasons that are supposed to support the conclusion are the *premises* of the argument.
- **Step 2**: Write each premise in a numbered line. Following the premises, write the conclusion after the word 'therefore'. Label each premise consecutively as 'P1', 'P2', and so on, and the conclusion as 'C'.
- **Step 3**: Determine whether the premises are supposed to provide logically conclusive or probable support for the conclusion of the argument. If the former, the argument is *deductive*, in which case, go to Step 4. If the latter, the argument is *inductive*, in which case, go to Step 6.
- **Step 4**: If the argument is deductive, determine whether the premises successfully provide logically conclusive support for the conclusion. If they do, the argument is *valid*, in which case, go to Step 5. If the premises fail to provide logically conclusive support for the conclusion, the argument is *invalid*, in which case, stop! It is not a good argument.
- **Step 5**: If the argument is valid, determine whether all the premises are in fact true. If all the premises are true, the argument is *sound*, in which case, stop! It is a good argument. If even one premise is false, the argument is *unsound*, in which case, stop! It is not a good argument.
- **Step 6**: If the argument is inductive, determine whether the premises successfully provide probable support for the conclusion. If they do, the argument is *strong*, in which case, go to Step 7. If the premises fail to provide probable support for the conclusion, the argument is *weak*, in which case, stop! It is not a good argument.
- **Step 7**: If the argument is strong, determine whether all the premises are in fact true. If all the premises are true, the argument is *cogent*, in which case, stop! It is

a good argument. If even one premise is false, the argument is *non-cogent*, in which case, stop! It is not a good argument.

Throughout this book, all the analyses and evaluations of key arguments in the scientific realism/antirealism debate in contemporary philosophy of science follow this generic decision procedure, so readers are invited to refer back to it whenever an analysis of a particular argument is at issue. Readers can also find the italicized argumentation terms along with their definitions in the Glossary at the end of this chapter.

Here is an example of how to use this generic decision procedure for analyzing and evaluating arguments in natural language. Take the following passage for example:

> Every field of inquiry deals with some subject matter: it studies something rather than nothing or everything. Thus it should be able to tell, at least roughly, what sort of objects it is concerned with and how its objects of study differ from those studied by other disciplines. [...] Evidently, what holds for all fields of inquiry also holds for a particular discipline such as the philosophy of science. Therefore, it belongs to the job description, so to speak, of the philosopher of science to tell us what that "thing" called science is (Mahner 2007, p. 515).

The first step is to figure out whether an argument is being made in this passage. In this passage, is evidence (or reasons) presented in support of a conclusion? The answer is yes. The word 'therefore' in the last sentence suggests that the conclusion of the argument in this passage is the following:

(C) Philosophers of science should be able to tell us what science is.

The author does not simply assert (C). Rather, the author provides evidence (or reasons) in support of (C). The reasons, which constitute the premises of this argument, are the following:

(P1) For any discipline or field of inquiry, practitioners in that field should be able to tell us what it is that they study.
(P2) In the discipline or field of inquiry known as philosophy of science, practitioners (namely, philosophers of science) study science.

Now that we have the parts of the argument, namely, the premises and the conclusion, the second step is to write the argument in canonical (or standard) form. When we put the premises and the conclusion together, we have the following argument in canonical (or standard) form:

(P1) For any discipline or field of inquiry, practitioners in that field should be able to tell us what it is that they study.
(P2) In the discipline or field of inquiry known as philosophy of science, practitioners (namely, philosophers of science) study science.

Therefore,

(C) Philosophers of science should be able to tell us what science is.

Now that we have the argument in canonical (or standard) form, the third step is to figure out what type of argument it is supposed to be: deductive or inductive. Are the premises of this argument supposed to provide *conclusive* or *probable* support for the conclusion? In this argument, the premises are supposed to provide conclusive support for the conclusion because they apply a general principle to a particular instance. Of course, what is true in general must be true in a particular instance that falls under the general principle. As the author of the passage puts it, "what holds for all fields of inquiry also holds for a particular discipline such as the philosophy of science" (Mahner 2007, p. 515). So this argument is supposed to be a deductive argument.

Given that this argument is intended to be a deductive argument, it can be either valid or invalid. The fourth step, then, is to figure out whether the argument is valid or invalid. If the premises were true, would they provide logically conclusive support for the conclusion? Again, the answer is yes. That is to say, the conclusion of this argument must be true if the premises are true. Again, what holds for all disciplines or fields of inquiry must hold for philosophy of science as well, given that philosophy of science is a discipline or field of inquiry.

Since this argument is valid, it can be either sound or unsound. The fifth step, then, is to figure out whether the premises of this argument are in fact true. Are (P1) and (P2) actually true? Naturally, this question is more difficult to answer than the questions about the type and logical form of the argument. Indeed, it seems fair to say that many philosophical disagreements are about the truth value of premises rather than the logical form of arguments. As we will see in Chaps. 4 and 5, philosophers of science often disagree about the premises of an argument, although they sometimes disagree about the logical form of arguments as well. In the case of the argument in question, however, there are good reasons to believe that (P1) is false. In particular, there seem to be plenty of counterexamples to (P1). Take psychology, for example. Psychologists study the mind even though it is difficult to say what the mind is exactly. There are philosophical disagreements about the nature of the mind: is it physical or non-physical? Is it identical to the brain? Is it like a computer? And so on. Despite these philosophical disagreements about the nature of the mind, psychologists can proceed with their studies of aspects of the mind, such as cognition, perception, reasoning, memory, and the like. In other words, psychologists study this thing called "the mind" even though they are unable to say what that thing called "the mind" is exactly because no one can, at least for now. Similarly, astronomers study planets even though it is difficult to say what a planet is exactly. There is an ongoing debate about the definition of 'planet', which has changed as recently as 2006, leading to the demotion of Pluto from the status of planet to the status of dwarf planet (deGrasse Tyson 2009, p. 119). Despite this debate about what a planet is, astronomers can proceed with their studies of the orbit, composition, and other properties of planets, such as Mars, Jupiter, and the like. In other words, astronomers study those celestial objects called "planets" even though they are unable to say what that thing called "a planet" is exactly because no one can, at least for now. Accordingly, there are disciplines or fields of inquiry, such as psychology and astronomy, in which practitioners are unable to tell us what it is that they study. So, the fact that philosophers of science study science does not necessarily mean that they would be able to tell us what science is exactly. If this is correct, then (P1) may not be true. Given this reason to think that (P1) is false, then, the argument in question, although valid, cannot be said to be a sound argument.

## 1.2   Just the Arguments

In addition to the pedagogical value of presenting arguments in canonical (or standard) form, so that we could analyze and evaluate them carefully, there is another important reason why there is a need for a book that takes an argumentation approach to the scientific realism/antirealism debate in contemporary philosophy of science, just as this book does. According to Michael Bruce and Steven Barbone (2011, p. 1):

> "Show me the argument" is the battle cry for philosophers. [...] When things become serious, one wants *just the arguments* (emphasis in original).

Given that the scientific realism/antirealism debate is a very serious thing, for it deals with serious questions about science that are of both theoretical and practical significance to all of us, it follows that we want *just the arguments* as far as this debate is concerned. This argument can be stated in canonical (or standard) form as follows:

(P1) When things become serious, we want *just the arguments*.
(P2) The scientific realism/antirealism debate is a serious thing.

Therefore,

(C) We want *just the arguments* in the scientific realism/antirealism debate.

To determine if this is a good argument, let us go through the generic decision procedure for the analysis and evaluation of arguments in natural language outlined above. We already have the parts of this argument, namely, the premises and the conclusion, stated in canonical (or standard) form. So we can skip to Step 3 and ask whether the premises of this argument are supposed to provide logically conclusive or probable support for the conclusion. In this argument, the premises are supposed to provide conclusive support for the conclusion because they apply a general principle to a particular instance. The general principle is that we want just the arguments when things become serious. Of course, what is true in general must be true in a particular instance that falls under the general principle. Accordingly, if we want just the arguments when things are serious, and the scientific realism/antirealism debate is a very serious thing, then it necessarily follows that we want just the arguments of the scientific realism/antirealism debate. Since this argument is supposed to be deductive, we can move on to Step 4 and ask whether it is valid or invalid.

This deductive argument is valid because the premises succeed in providing logically conclusive support for the conclusion. In particular, if we want *just the arguments* when things are serious, and the scientific realism/antirealism debate is a very serious thing, then it follows logically that we want *just the arguments* in the scientific realism/antirealism debate. Having determined that this argument is valid, we can move on to Step 5 and ask whether the premises are in fact true. Is this argument sound?

For the sake of argument, let us grant that Bruce and Barbone (2011) are right about (P1); after all, few philosophers, if any, would deny that "philosophers are primarily concerned with arguments" (Harrell 2016, p. 7). Now the question is whether (P2) is true. I believe that it is, and I have reasons in support of my belief that (P2) is true. As mentioned above, I think that the contemporary scientific realism/antirealism debate is a very serious thing, for it deals with serious questions about science that are of enormous theoretical and practical significance for all of us. As Brad Wray (2018, p. 1) points out, one of the central questions in the scientific realism/antirealism debate is this: "Do we have adequate grounds for believing that our theories are true or approximately true with respect to what they say about unobservable entities and processes?" The "unobservable entities and processes" that Wray is talking about here are the theoretical posits of science, which include theoretical entities, such as neutrinos and genes, as well as theoretical processes, such as natural selection and continental drift. Since science informs many of our decisions nowadays, especially at the social and political levels, whether one takes a realist or an antirealist attitude with respect to the theoretical posits of science will in turn inform or influence one's decisions about social and political issues that could affect many people, not just oneself. To illustrate, take the public debate concerning the legalization of marijuana for example. One of the arguments made in this debate is that marijuana use should not be legalized because smoking marijuana could have profound effects on brain functions, such as working memory and executive function. This is because the main active ingredient of cannabis, THC, acts as a neurotransmitter throughout the nervous system. More specifically (National Institute on Drug Abuse 2019, p. 17):

> *Endogenous cannabinoids* such as anandamide function as *neurotransmitters* because they send chemical messages between nerve cells (*neurons*) throughout the nervous system. They affect brain areas that influence pleasure, memory, thinking, concentration, movement, coordination, and sensory and time perception. Because of this similarity, THC is able to attach to molecules called *cannabinoid receptors* on neurons in these brain areas and activate them, disrupting various mental and physical functions and causing the effects described earlier. The neural communication network that uses these cannabinoid neurotransmitters, known as the *endocannabinoid system,* plays a critical role in the nervous system's normal functioning, so interfering with it can have profound effects (emphasis in original).

Now, for the most part, scientific realists tend to think that we have good reasons to believe what our best scientific theories say about the brain, which is why they believe that theoretical posits, such as molecules, neurons, neurotransmitters, and receptors, have roughly the properties that our best scientific theories about the brain say they do. On the other hand, antirealists tend to think that we do not have good reasons to believe what our best scientific theories say about theoretical posits, which is why they suspend belief about the existence of theoretical posits, such as molecules, neurons, neurotransmitters, and receptors. Clearly, then, whether one takes a realist or an antirealist attitude toward brain science or neuroscience will inform or influence (perhaps even tacitly) one's decisions about the legalization of marijuana. For if one does not think that we have good reasons to believe in the theoretical posits of neuroscience, such as neurons, neurotransmitters, and receptors, and in what our best theories say about these theoretical posits, then it is difficult to see how one can believe that smoking marijuana could have profound effects on brain function. After all, neurons, neurotransmitters, and receptors are supposed to be the mechanisms by which cannabis can have an effect on the brain. If one does not believe in the existence of the mechanisms by which cannabis can have an effect on the brain, and in what our best neuroscience says about those mechanisms, it is difficult to see how one can believe that THC can attach to receptors on neurons in the brain and thereby disrupt brain functions.

As another example, take the public debate concerning social distancing measures designed to prevent the spread of the infectious disease known as Coronavirus Disease 2019 (COVID-19), which is believed to be caused by the Severe Acute Respiratory Syndrome Coronavirus 2 (SARS-CoV-2). According to the Centers for Disease Control and Prevention (CDC), social distancing is the practice of keeping space or a physical distance of at least 6 ft (or 2 m) between oneself and other people when one leaves one's home. If we want to prevent the spread of COVID-19, then practicing social distancing makes sense because of how we think the virus spreads. From what we know about the new coronavirus so far, it spreads from one person to another by means of respiratory droplets that are produced when an infected person coughs or sneezes around other people. An infected person's cough or sneeze produces respiratory droplets that can land in the mouths or noses of people who are nearby. These respiratory droplets can also land on the hands of nearby people who will then touch their own mouths or noses, and thereby provide a way for the virus to enter their bodies. Finally, the virus could also be inhaled into the lungs.

Now, as mentioned above, scientific realists generally think that we have good reasons to believe what our best scientific theories say about infectious, disease-causing agents known as "germs" or "pathogens," which is why they believe that theoretical posits, such as germs and viruses, have roughly the properties that our best scientific theories say they do. On the other hand, antirealists generally think that we do not have good reasons to believe what our best scientific theories say about theoretical posits, which is why they suspend belief about the existence of theoretical posits, such as germs and viruses. Clearly, then, whether one takes a realist or an antirealist attitude toward epidemiology and virology will inform or influence (perhaps even tacitly) one's decisions about what measures, if any, should be taken in order to prevent the spread of an infectious disease like COVID-19. For if one does not think that we have good reasons to believe in the theoretical posits of epidemiology and virology, such as viruses and respiratory droplets, and in what our best theories say about these theoretical posits, then it is difficult to see how one can believe that social distancing could have an effect on the spread of viral infections like COVID-19. After all, the virus and respiratory droplets are supposed to be the mechanisms by which the infectious disease can spread from one person to another. If one does not believe in the existence of the mechanisms by which the infectious disease can spread, and in what our best medical science says about those mechanisms, it is difficult to see how one can believe that social distancing can have any effect on the spread of this infectious disease.[3]

---

[3] Dana Tulodziecki (2016) discusses the germ theory of disease as a case study for the scientific realism/antirealism debate in contemporary philosophy of science.

The point of these examples is to illustrate the seriousness of the scientific realism/antirealism debate in contemporary philosophy of science. These are just two examples of how science can inform or influence decisions about public policies, but there are many other examples. Take, for example, the public debates over policies to combat climate change, the safety of genetically modified food, and the move to alternative, renewable, or "green" energy sources, to mention just a few. Whether one believes the relevant science or not, then, would surely have some effect on one's decisions and any public policies one would be willing to endorse. Accordingly, if the scientific realism/antirealism debate in contemporary philosophy of science is a very serious thing, as I have argued, then (P2) is true. Granted that (P1) is true as well, since "argument is the heart of philosophy" (Cohen 2004, p. 117) and argumentation is "the most basic philosophical technique" (Martin 2017, p. IX), the deductive argument sketched above can be said to be sound. This sound argument supports the conclusion that we want *just the arguments* as far as the scientific realism/antirealism debate in contemporary philosophy of science is concerned, and this book is aimed at satisfying precisely this want.

In my evaluation of the argument for the conclusion that we want just the arguments in the scientific realism/antirealism debate above, I said that few philosophers, if any, would object to (P1), namely, the premise that we want just the arguments when things become serious. This is because "philosophers are primarily concerned with arguments" (Harrell 2016, p. 7). In general, then, the main key to successful argumentation is to argue from premises that one's interlocutor or audience is likely to accept (or at least can be reasonably expected to accept). For, as Trudy Govier (2010, p. 25) puts it:

> When you use an argument, you are trying to *rationally persuade* others of the claim that is your conclusion. You are trying to *convince* them, by evidence or reasons stated in your premises, that your conclusion claim is correct and you are offering the premises in an attempt to rationally persuade them. In effect, *you are asking your audience to accept your premises and to reason from those premises to your conclusion* (emphasis added).

If your interlocutors do not accept your premises, they will not reason from your premises to your conclusion, and thus your attempt to convince them, by evidence or reasons stated in your premises, that your conclusion is correct will end in failure. If you want your attempts to convince others, by evidence or reasons stated in your premises, that your conclusion is correct, to be successful, you need to argue from premises that your interlocutors are likely to accept (or can be reasonably expected to accept). As we will see in subsequent chapters, scientific realists often argue from premises that antirealists do not (or are unlikely to) accept, whereas antirealists argue from premises that realists do not (or are unlikely to) accept. This fact about the scientific realism/antirealism debate in contemporary philosophy of science has given some philosophers of science the impression that the debate is intractable. For example, Allison Wylie (1986) argues that the scientific realism/antirealism debate "persists because the most sophisticated positions in either side now incorporate *self-justifying conceptions of the aim of philosophy* and of the standards of adequacy appropriate for judging philosophical theories of science" (emphasis added). In other words, scientific realists are arguing from meta-philosophical premises (that is, premises about what counts as a good philosophical theory of science) that antirealists cannot accept, whereas antirealists are arguing from meta-philosophical premises (that is, premises about what counts as a good philosophical theory of science) that scientific realists cannot accept. If Wylie is right about this, then it is no wonder that the scientific realism/antirealism debate seems to be intractable.[4]

I think that taking an argumentation approach might help us break through this apparent impasse in the scientific realism/antirealism debate. Keeping in mind that the main key to successful argumentation, that is, the giving of reasons (or evidence) that manage to rationally persuade an audience of the conclusion of an argument, is to argue from premises that one's interlocutors are likely to accept (or can be reasonably expected to accept), I will analyze and evaluate what I take to be key arguments in the scientific realism/antirealism debate in such a way that both the acceptable and the unacceptable premises to either scientific realists or antirealists are clear to the reader. Subsequently, I will attempt to give arguments for my own position in the scientific realism/antirealism debate, namely, Relative Realism, that proceed from premises that both scientific realists and antirealists can accept, or so I would argue.

Before we get to that, however, I will introduce the contemporary scientific realism/antirealism debate in more detail in the next chapter (Chap. 2), and I will further explain what is at stake in this debate. In Chap. 3, I will survey what I take to be key positions in the scientific realism/antirealism debate in contemporary philosophy of science. Some of these positions are realist positions, broadly speaking, whereas others are antirealist positions, broadly speaking. In Chap. 4, I will analyze and evaluate what I take to be key arguments for scientific realism using the argumentation approach described above. In Chap. 5, I will analyze and evaluate what I take to be key arguments against scientific realism (or for antirealism) using the argumentation approach described above. Finally, in Chap. 6, I will discuss my own brand of scientific realism, namely, Relative Realism. I take Relative Realism to be a middle ground position between scientific realism and antirealism. I have proposed this view for the first time in a paper published in *International Studies in the Philosophy of Science* in 2013. But I will develop this position in much more detail, as well as advance novel arguments for it, in Chap. 6 of this book.

## 1.3  Summary

The scientific realism/antirealism debate in contemporary philosophy of science is a very serious thing, which is why we need *just the arguments* as far as this debate is concerned. When it comes to arguments, we can broadly distinguish between two kinds: deductive arguments and non-deductive (or inductive) arguments. In deductive argumentation, a premise (or premises) purports to provide logically conclusive

---

[4] Paul Dicken (2016, Ch. 3) provides an extensive discussion of the apparent intractability of the scientific realism/antirealism debate, starting with Arthur Fine's (1984).

support for a conclusion. Deductive arguments in which the premises succeed in providing logically conclusive support for the conclusion are said to be *valid* arguments; otherwise, they are said to be *invalid* arguments. A deductive argument that is valid and all of its premises are true is said to be a *sound* argument; if even a single premise is false, it is said to be an *unsound* argument. In non-deductive or inductive argumentation, a premise (or premises) purports to provide probable support for a conclusion. Inductive arguments in which the premises succeed in providing probable support for the conclusion are said to be *strong* arguments; otherwise, they are said to be *weak* arguments. An inductive argument that is strong and all of its premises are true is said to be a *cogent* argument; if even a single premise is false, it is said to be a *non-cogent* argument. The main key to successful argumentation is to argue from premises that one's interlocutor or audience is likely to accept (or at least can be reasonably expected to accept).

# Glossary

**Antirealism** An agnostic or skeptical attitude toward the theoretical posits (that is, unobservables) of scientific theories. Antirealism comes in different varieties, such as Constructive Empiricism (see Chap. 3, Sect. 3.3) and Instrumentalism (see Chap. 3, Sect. 3.2).

**Approximate truth** Closeness to the truth or truthlikeness. To say that a theory is approximately true is to say that it is close to the truth. According to some scientific realists, approximate truth is the aim of science. (See Chap. 2, Sect. 2.1.)

**Argument** A set of statements in which some (at least one statement called a premise) purport to provide logical support (either deductive or inductive) for another statement (namely, the conclusion). (See Chap. 1, Sect. 1.1.)

**Canonical form** A method of representing arguments where each premise is written on a separate, numbered line, followed by the conclusion (also known as "standard form"). (See Chap. 1, Sect. 1.1.)

**Cogent argument** A strong argument with all true premises. (See Chap. 1, Sect. 1.1.)

**Conclusion** The statement in an argument that the premises purport to support. (See Chap. 1, Sect. 1.1.)

**Deduction** A form of argumentation in which the premises purport to provide logically conclusive support for the conclusion. (See Chap. 1, Sect. 1.1.)

**Fallacious argument** An argument whose premises fail to provide either conclusive or probable support for its conclusion (see also *invalid argument* and *weak argument*). (See Chap. 2, Sect. 2.2.)

**Induction** A form of argumentation in which the premises purport to provide probable support for the conclusion. (See Chap. 1, Sect. 1.1.)

**Invalid argument** A deductive argument in which the premises purport but fail to provide logically conclusive support for the conclusion. (See Chap. 1, Sect. 1.1.)

**Modus ponens** A form of argument with a conditional premise, a premise that asserts the antecedent of the conditional premise, and a conclusion that asserts the consequent of the conditional premise. That is, "if *A*, then *B*, *A*; therefore, *B*," where *A* and *B* stand for statements. *Modus ponens* is a valid form of inference, and so an argument in natural language that takes this logical form is valid. On the other hand, the following logical form is invalid: "if *A*, then *B*, *B*; therefore, *A*." It is known as the fallacy of affirming the consequent. (See Chap. 4, Sect. 4.1.)

**Modus tollens** A form of argument with a conditional premise, a premise that denies the consequent of the conditional premise, and a conclusion that denies the antecedent of the conditional premise. That is, "if *A*, then *B*, not *B*; therefore, not *A*," where *A* and *B* stand for statements. *Modus tollens* is a valid form of inference, and so an argument in natural language that takes this logical form is valid. On the other hand, the following logical form is invalid: "if *A*, then *B*, not *A*; therefore, not *B*." It is known as the fallacy of denying the antecedent. (See Chap. 5, Sect. 5.1.)

**Scientific realism** An epistemically positive attitude toward those aspects of scientific theories that are worthy of belief. Scientific realism comes in different varieties, such as Explanationist Realism (see Chap. 3, Sect. 3.1), Entity Realism (see Chap. 3, Sect. 3.4), Structural Realism (see Chap. 3, Sect. 3.5), and Relative Realism (see Chap. 6, Sect. 6.1).

**Sound argument** A valid argument with all true premises. (See Chap. 1, Sect. 1.1.)

**Strong argument** A non-deductive (or inductive) argument in which the premises successfully provide probable support for the conclusion. (See Chap. 1, Sect. 1.1.)

**Premise** A statement in an argument that purports to support the conclusion of that argument. (See Chap. 1, Sect. 1.1.)

**Valid argument** A deductive argument in which the premises successfully provide logically conclusive support for the conclusion. (See Chap. 1, Sect. 1.1.)

**Weak argument** A non-deductive (or inductive) argument in which the premises purport but fail to provide probable support for the conclusion. (See Chap. 1, Sect. 1.1.)

## References and Further Readings

Berenstain, N., & Ladyman, J. (2012). Ontic structural realism and modality. In E. Landry & D. Rickles (Eds.), *Structural realism: Structure, object, and causality* (pp. 149–168). Dordrecht: Springer.

Bourget, D., & Chalmers, D. J. (2014). What do philosophers believe? *Philosophical Studies, 170*(3), 465–500.

Bruce, M., & Barbone, S. (2011). Introduction: Show me the argument. In M. Bruce & S. Barbone (Eds.), *Just the arguments: 100 of the most important arguments in western philosophy* (pp. 1–6). Oxford: Blackwell.

Chakravartty, A. (2017). Scientific realism. In E. N. Zalta (Ed.), *The Stanford encyclopedia of philosophy*. Summer 2017 Edition. https://plato.stanford.edu/archives/sum2017/entries/scientific-realism/.

Cohen, D. (2004). *Arguments and metaphors in philosophy*. New York: University Press of America.

deGrasse Tyson, N. (2009). *The Pluto files: The rise and fall of America's favorite planet*. New York: W. W. Norton & Co.

Dicken, P. (2016). *A critical introduction to scientific realism*. London: Bloomsbury.

Fine, A. (1984). In S. Realism & e. J. Leplin (Eds.), *The natural ontological attitude* (pp. 83–107). Berkeley: University of California Press.

Govier, T. (2010). *A practical study of argument* (7th ed.). Belmont, CA: Wadsworth.

Harrell, M. (2016). *What is the argument? An introduction to philosophical argument and analysis*. Cambridge, MA: The MIT Press.

Mahner, M. (2007). Demarcating science from nonscience. In T. A. F. Kuipers (Ed.), *General philosophy of science: Focal issues* (pp. 515–576). Amsterdam: Elsevier.

Martin, R. M. (2017). *For the sake of argument: How to do philosophy*. Peterborough, Ontario: Broadview Press.

Mizrahi, M. (2018). Introduction. In M. Mizrahi (Ed.), *The Kuhnian image of science: Time for a decisive transformation?* (pp. 1–22). London: Rowman & Littlefield.

Mizrahi, M. (2020). The case study method in philosophy of science: An empirical study. *Perspectives on Science, 28*(1), 63–88.

National Institute on Drug Abuse. (2019). Marijuana research report series. In J. J. Gonzalez III & M. P. McGee (Eds.), *Legal Marijuana: Perspectives on public benefits, risks and policy approaches* (pp. 15–19). Jefferson, NC: McFarland and Co.

Psillos, S. (1999). *Scientific realism: How science tracks truth*. London: Routledge.

Sankey, H. (2008). *Scientific realism and the rationality of science*. Hampshire: Ashgate.

Schurz, G. (2019). *Hume's problem solved: The optimality of meta-induction*. Cambridge, MA: The MIT Press.

Tulodziecki, D. (2016). From zymes to germs: discarding the realist/anti-realist framework. In R. Scholl & T. Sauer (Eds.), *The philosophy of historical case studies* (pp. 265–284). Basel: Springer.

Wray, B. K. (2018). *Resisting scientific realism*. Cambridge: Cambridge University Press.

Wylie, A. (1986). Arguments for scientific realism: The ascending spiral. *American Philosophical Quarterly, 23*(3), 287–297.

# Chapter 2
# Realism Versus Antirealism
# in Contemporary Philosophy of Science

**Abstract** The scientific realism/antirealism debate in contemporary philosophy of science is about theoretical knowledge, that is, knowledge that is supposed to be about so-called "unobservables." This includes theoretical entities, such as neutrinos and genes, as well as theoretical processes, such as natural selection and continental drift. Is theoretical knowledge in science possible? In general, scientific realists tend to think that science can (and does) yield theoretical knowledge, whereas antirealists tend to think that science cannot (and does not) yield theoretical knowledge (that is, knowledge about unobservables). To put it another way, scientific realists tend to argue that we have good reasons to believe that our best scientific theories are approximately true because, if they were not even approximately true, they would not be able to explain and predict natural phenomena with such impressive accuracy. On the other hand, antirealists tend to argue that the impressive success of our best scientific theories does not warrant belief in the approximate truth of our best scientific theories. This is because the history of science is a graveyard of theories that were once successful but were later discarded.

**Keywords** Antirealism · Approximate truth · Case study · Epistemic dimension (stance or thesis) of scientific realism · Hasty generalization · Metaphysical dimension (stance or thesis) of scientific realism · Observable/unobservable distinction · Scientific realism · Selectivist turn · Semantic dimension (stance or thesis) of scientific realism · Theoretical knowledge

The results of a Pew Research poll from 2015 show that there are big differences of opinion on scientific issues between professional scientists and the general public. For example, 98% of scientists "connected to the American Association for the Advancement of Science (AAAS)" say that humans have evolved over time, whereas only 65% of U.S. adults accept that humans have evolved over time. Likewise, 87% of AAAS scientists say that climate change is mostly due to human activity, but only 50% of U.S. adults accept that climate change is mostly due to human activity. Why are there such big differences of opinion on scientific issues between professional scientists and the general public? Do scientists have good reasons to believe in what their best scientific theories say about evolution and climate change? If so, what are those reasons? And why is the general public

© Springer Nature Switzerland AG 2020
M. Mizrahi, *The Relativity of Theory*, Synthese Library 431,
https://doi.org/10.1007/978-3-030-58047-6_2

suspicious of these scientific theories? Are there good reasons to suspend belief in what the best scientific theories say about evolution and climate change? If so, what are those reasons? These questions are at the core of the scientific realism/antirealism debate in contemporary philosophy of science.

The scientific realism/antirealism debate in contemporary philosophy of science is primarily about the epistemic status of scientific theories. As Brad Wray (2018, p. 1) puts it, one of the central questions in the contemporary scientific realism/antirealism debate is this: "Do we have adequate grounds for believing that our theories are true or approximately true with respect to what they say about unobservable entities and processes?"[1] That is to say, scientific theories make theoretical claims about the world around us. For example, according to the modern theory of evolution in biology, all living organisms on Earth are related by descent with modification from common ancestors. According to the theory of anthropogenic climate change in climate science, the main cause of global climate change is the greenhouse gases emitted by human activities. Should we believe what these scientific theories say about the world around us? After all, these theories make claims about entities and processes that cannot be observed with the naked eye. For example, the modern theory of evolution in biology makes claims about processes that occur on very large scales and over very long periods of time, such as genetic drift and natural selection, and about very small entities, such as genes and DNA. Such processes and entities cannot be directly observed with the naked eye. Likewise, the theory of anthropogenic climate change in climate science makes claims about processes that occur on very large scales and over very long periods of time, such as infrared energy transfer and the greenhouse effect, and about very small entities, such as photons and electromagnetic waves. Again, such processes and entities cannot be directly observed with the naked eye. Rather, such processes and entities are theoretical posits insofar as they are posited in order to explain natural phenomena that we can observe. In the case of the theory of evolution, for example, the existence of processes like natural selection and entities like genes is postulated to explain biodiversity and speciation, that is, how new species come into being. Likewise, in the case of the theory of anthropogenic climate change in climate science, the existence of processes like the transfer of infrared energy and the heat-trapping properties of greenhouse gases, such as carbon dioxide and methane, is postulated to explain global warming trends. Should we believe in the existence of these processes and entities that cannot be directly observed with the naked eye but rather are postulated in order to explain observed phenomena? This question keeps philosophers of science up at night.[2]

---

[1] As Darrell Rowbottom (2019) points out, there is also an axiological dimension to the scientific realism/antirealism debate in contemporary philosophy of science. This axiological dimension is about the goals or aims of science. I will say more about the axiological dimension in Chap. 5 (Sect. 5.5).

[2] Johanna Wolff (2019) points out that some philosophers of science prefer to not take sides in the scientific realism/antirealism debate. She calls these philosophers of science "quietists."

## 2.1   The Three Dimensions of Scientific Realism

Typically, scientific realists tend to believe in the existence of the entities and pro-cesses posited by our best scientific theories. As we will see in Chap. 3, while there are several different positions that fall under the category of scientific realism, broadly speaking, most participants in the scientific realism/antirealism debate in contemporary philosophy of science generally associate scientific realism with one (or more) of the following theses (or stances):

1. The *metaphysical stance* asserts that the world has a definite and mind-independent natural-kind structure.
2. The *semantic stance* takes scientific theories at face-value, seeing them as truth-conditioned descriptions of their intended domain, both observable and unob-servable. Hence, they are capable of being true or false. Theoretical assertions are not reducible to claims about the behaviour of observables, nor are they merely instrumental devices for establishing connections between observables. The theoretical terms featuring in theories have putative factual reference. So, if scientific theories are true, the unobservable entities they posit populate the world.
3. The *epistemic stance* regards mature and predictively successful scientific theo-ries as well-confirmed and approximately true of the world. So, the entities pos-ited by them, or, at any rate, entities very similar to those posited, do inhabit the world (Psillos 1999, p. xvii).[3]

As Stathis Psillos and Emma Ruttkamp-Bloem (2017) point out, parties to the sci-entific realism/antirealism debate in contemporary philosophy of science generally agree "that scientific realism has three dimensions (Psillos 1999) or stances (Chakravartty 2007); a metaphysical, semantic, and an epistemic dimension."

The metaphysical thesis (or stance or dimension) of scientific realism is sup-posed to capture the idea that there are things out there in the world for scientists to discover and that those things out there in the world are independent of the human minds that study them. That is to say, when scientists announce that they have made a discovery, they are not merely talking about the figments of their own imagina-tions. Rather, they are talking about real things that exist in the world, even though those things are unobservable, that is, they cannot be directly observed with the naked eye. For example, when scientists working with the Large Hadron Collider (LHC) in Geneva announced that they had found the elementary particle known as the Higgs boson in 2012, they were not merely making things up. Rather, they were talking about a real thing that exists in nature, even though this real thing, namely, the Higgs boson, is unobservable, that is, it takes sophisticated scientific instru-ments, such as particle accelerators, colliders, and the like, to detect elementary particles like the Higgs boson. One of the main questions in the scientific realism/antirealism debate in contemporary philosophy of science is when, and under what

---

[3] Anjan Chakravartty (2007, p. 10) provides a useful classification of realist and antirealist views about science.

circumstances (if any), we are justified in believing that what our best scientific theories say about unobservables (for example, about elementary particles, such as the Higgs boson) is likely true or approximately true.

This last point ties in with the semantic thesis (or stance or dimension) of scientific realism mentioned above. Scientific realists (and even some antirealists, as we will see in Chap. 3) think that scientific theories can be true or false. For example, the theoretical statement "The Higgs boson has no electric charge" is either true or false. Currently, our best theory of elementary particles, which is known as the Standard Model of particle physics, tells us that this theoretical statement is true, that is, that the Higgs boson has no electric charge. Similarly, the theoretical statement "SARS-CoV-2 is the virus that causes COVID-19" is either true or false. Our best current evidence points to this theoretical statement being true, that is, that SARS-CoV-2 is the virus that causes COVID-19. When scientific realists talk about the truth of a theoretical statement, they typically understand it to mean "correspondence with reality" (Psillos 1999, p. 248). That is to say, a theoretical statement is true just in case what it says about reality corresponds with reality. As Howard Sankey (2008, p. 16) puts it, on the so-called correspondence theory of truth, "truth is a relation of correspondence that obtains in virtue of the world in fact being the way that it is said to be." For example, if the theoretical statement "SARS-CoV-2 is the virus that causes COVID-19" is true, then it is in fact the case that SARS-CoV-2 is the virus that causes COVID-19. If it is not the case that SARS-CoV-2 is the virus that causes COVID-19, then the theoretical statement "SARS-CoV-2 is the virus that causes COVID-19" is false. As we will see in Chap. 3, some antirealists deny this semantic thesis (or stance or dimension) of scientific realism. Instead, they argue that scientific theories should be construed as instruments or tools for prediction, not as statements that have truth values (either true or false).

Of course, if theoretical statements can be either true or false, the question is how do we know that they are true (or false). As mentioned above, this is one of the key questions in the scientific realism/antirealism debate in contemporary philosophy of science. Some scientific realists argue that we sometimes do know (or have good reasons to believe) that our best scientific theories are likely true (or approximately true). Those realists are committed to the epistemic thesis (or stance or dimension) of scientific realism mentioned above. These scientific realists argue that the fact that our best scientific theories make predictions that are borne out by observation and experimentation provides a good reason to believe that our best scientific theories are approximately true. This realist argument will be analyzed and evaluated in Chap. 4 (see Sect. 4.1). On the other hand, some antirealists argue against predictive success being a sign of the (approximate) truth of our best scientific theories on the grounds that the history of science provides examples of theories that were predictively successful (that is, theories that made predictions that were borne out by observation and experimentation), but that were later discarded or abandoned. This antirealist argument will be analyzed and evaluated in Chap. 5 (see Sect. 5.1).

In that respect, we should note the use of the terms 'approximate' and 'approximately' with respect to the truth of scientific theories as they are used in the epistemic thesis (or stance or dimension) of scientific realism. Most scientific realists and some antirealists agree that, strictly speaking, scientific theories, even the best ones, are not entirely true. As Anjan Chakravartty (2017) puts it, "it is widely held, not least by realists, that even many of our best scientific theories are likely false, strictly speaking." The fact that a scientific theory is not entirely true, however, does not necessarily mean that it is completely false. In other words, a scientific theory is not a monolithic whole; it can have some false parts and some true parts. Accordingly, a scientific theory can be said to be approximately true in the sense of being close to the truth to the extent that it has some true parts, but also not fully true to the extent that it has a few false parts as well. As Philip Kitcher (2002, p. 388) puts it, "It doesn't follow from the fact that a past theory isn't completely true that every part of that theory is false."

It should also be noted, however, that explicating the notion of approximate truth is notoriously difficult. Karl Popper's (1972) attempt to formalize the notion of approximate truth or verisimilitude was shown to be problematic (see, for example, Miller 1974 and Tichý 1974). Other formal approaches, such as the similarity approach (see, for example, Niiniluoto 1998) and the type hierarchy approach (see, for example, Aronson et al. 1994) also suffer from technical problems (see, for example, Psillos 1999). For these reasons, scientific realists have tried to explicate approximate truth in non-formal, qualitative terms (see, for example, Leplin 1981 and Boyd 1990). For example, it has been suggested that $T_2$ is more approximately true than its predecessor $T_1$ if $T_1$ can be described as a "limiting case" of $T_2$ (see, for example, Post 1971). But there are problems with these informal approaches as well (Chakravartty 2010). Here, we need not get into the technical details of these formal and informal accounts of approximate truth or verisimilitude. For our present purposes, we can simply follow philosophers of science like Stathis Psillos (1999, p. 278) and work with an "intuitive understanding of approximate truth, or of truth-likeness," while accepting that "there is no relevant need for a formal introduction of the predicate 'is approximately true', or the predicate 'is truth-like'." We will return to the notion of approximate truth in Chap. 6, however, where I will offer a definition that fits with the middle ground position I call "Relative Realism."

As we will see in Chap. 3, positions that fall under the broad category of scientific realism may vary along the metaphysical, semantic, and epistemological dimensions mentioned above. In general, however, scientific realists seek to identify the content of scientific theories that, in their view, warrants belief. As Anjan Chakravartty (2017) puts it, "Scientific realism is a positive epistemic attitude toward the content of our best theories and models, recommending belief in both observable and unobservable aspects of the world described by the sciences." As we will see in Chap. 3 as well, just like scientific realism, antirealism also comes in different varieties. Nevertheless, it is fair to say that scientific antirealists are generally suspicious of theoretical claims about theoretical posits that cannot be observed directly, that is, "unobservables" or entities, processes, and events that cannot be observed with the naked eye or without the aid of scientific instruments. In that respect, it is important to note that antirealists are *not* anti-science. They are generally willing to grant that science yields knowledge about the world around us. For example, antirealists would grant that biologists know a great deal about the different species of plants and animals that inhabit the Earth. What antirealists would question, however, is whether biologists know that the mechanisms of natural selection, genetic drift, and the like, are responsible for the biodiversity observed on planet Earth. Likewise, antirealists would admit that climatologists know a great deal about the atmosphere of the Earth. What antirealists would question, however, is whether climatologists know that the mechanisms of infrared energy transfer, heat-trapping, and the like, are responsible for the warming trends observed on planet Earth.

Generally speaking, then, antirealists tend to distinguish between what is observable and what is unobservable. Anything that can be directly observed with the naked eye counts as observable. This includes things like goats, geysers, and glaciers. Anything that cannot be directly observed with the naked eye, but rather requires the use of scientific instruments, such as telescopes and microscopes, in order to be detected, counts as unobservable. This includes things like gluons, genes, and gravitational waves. As far as the scientific realism/antirealism debate in contemporary philosophy of science is concerned, both scientific realists and antirealists are particularly interested in claims to *theoretical* knowledge, which are claims about unobservable entities, processes, or events. In other words, the sort of scientific knowledge that is at stake in the scientific realism/antirealism debate in contemporary philosophy of science, then, is theoretical knowledge. For antirealists tend to be suspicious of theoretical knowledge, or knowledge about unobservables, not observational knowledge, or knowledge about observables, in science.

Accordingly, the scientific realism/antirealism debate in contemporary philosophy of science is about theoretical knowledge, that is, knowledge that is supposed to be about theoretical entities, such as neutrinos and genes, as well as theoretical processes, such as natural selection and continental drift. Is theoretical knowledge in science possible? In general, scientific realists tend to argue that science can (and does) yield theoretical knowledge, whereas antirealists tend to argue that science cannot (and does not) yield theoretical knowledge (that is, knowledge of unobservables). To put it another way, scientific realists tend to argue that we have good reasons to believe that our best scientific theories are approximately true because, if they were not even approximately true, they would not be able to explain and predict natural phenomena with such impressive accuracy. On the other hand, antirealists tend to argue that the impressive empirical success of our best scientific theories does not warrant belief in the approximate truth of our best scientific theories. This is because the history of science is a graveyard of theories that were once successful but were later discarded. These arguments will be analyzed and evaluated in Chap. 4 (see Sect. 4.1) and Chap. 5 (see Sect. 5.1), respectively.

Before we begin our survey of key positions and arguments in the scientific realism/antirealism debate in contemporary philosophy of science, it is important to note that the focus of this book is the *contemporary* debate. As Psillos and Ruttkamp-Bloem (2017, p. 3190) observe, while scientific realism used to include at least three theses: "Theoretical terms refer to unobservable entities; … theories are (approximately) true; and … there is referential continuity in theory change," contemporary scientific realists tend to be more selective about the content of scientific theories that they identify as worthy of belief (see also Psillos 2018). That is to say, the scientific realism/antirealism debate used to be dominated by discussions of the notion of convergence on the truth, that is, that "mature theories are converging on the truth because they all are referring to the same things" (Laymon 1984, p. 121, cf. Laudan 1981). Nowadays, however, scientific realists rarely use the notions of convergence on the truth and reference of theoretical terms in general.[4] Instead, they try to be more selective about the parts of scientific theories that warrant a realist commitment in their view. In that respect, the scientific realism/antirealism debate in contemporary philosophy of science has taken what might be called a "selectivist turn" insofar as contemporary scientific realists are more selective about the aspects of science they are willing to be realists about than previous realists were. This selectivist turn has given rise to the following selective realist positions: Explanationist Realism, Entity Realism, and Structural Realism.[5]

Rather than having a realist commitment to the theoretical posits of our best scientific theories in general, explanationist realists recommend limiting one's realist commitments to those theoretical posits that are responsible for, or best explain, the predictive success of our best scientific theories. Entity realists recommend reserving one's realist attitudes to those theoretical posits that can be causally manipulated as well as enable efficacious interventions in nature. For structural realists, the parts of our best scientific theories we should be realists about are not theoretical entities, processes, or events per se, but rather the structures posited by such theories. These positions will be discussed in Chap. 3 (see Sects. 3.1, 3.4, and 3.5, respectively).

## 2.2 "Just Say No" (to Case Studies)

As mentioned in Chap. 1, rather than devoting an entire book to a defense of scientific realism (of some variety or another) or antirealism (of some variety or another) by describing in detail case studies from the history of science that are alleged to support the former or the latter, this book takes an argumentation approach to the scientific realism/antirealism debate in contemporary philosophy of science. In addition to the advantages of this argumentation approach discussed in Chap. 1, this approach also shows why providing detailed descriptions of a few case studies from the history of science is not likely to be a successful argumentative strategy in the contemporary scientific realism/antirealism debate.

As we have seen in Chap. 1 (see Sect. 1.1), a strong argument is an inductive argument whose premises successfully provide probable support for its conclusion.

---

[4] This is not to say that there is no mention of the notion of reference in the contemporary scientific realism/antirealism debate at all. For example, Paul Needham (2018) discusses whether 'water' preserved its extension in chemistry.

[5] Milena Ivanova (2013) provides a useful discussion of some of the historical actors in the scientific realism/antirealism debate, such as Pierre Duhem, Henri Poincaré, and Ernst Mach.

In other words, an inductive argument is strong when its premises, if true, make the conclusion more probable or likely to be true. Now, according to the *Oxford English Dictionary*, a case study is "a *particular* instance of something used or analyzed in order to illustrate a *thesis* or principle" (emphasis added). In philosophy of science, then, case studies are "*particular*, detailed descriptions of scientific activity" (Currie 2015, p. 553). The question, then, is this: Can a particular, detailed description of a scientific activity (that is, a case study) provide evidential support (either logically conclusive, as in deductive arguments, or probable, as in inductive arguments) for a thesis or principle in the scientific realism/antirealism debate in philosophy of science? Or, as Michael Bishop and J. D. Trout (2002, p. S204) ask, "How much support does a single case study (or even a number of case studies) provide a general principle about the nature of science?"

Recall from our discussion above that theses and/or principles in the scientific realism/antirealism debate in philosophy of science are supposed to be rather *general*, not particular. For, as we have seen, scientific realism is generally characterized as an "epistemically positive attitude toward those aspects of theories that are most worthy of epistemic commitment" (Chakravartty 2017). In that respect, scientific realism (of one variety or another) is meant to be a general thesis or "a general principle about the nature of science," not just a claim about a particular science, a particular scientific theory, or a particular theoretical posit. The problem is that a particular, detailed description of a scientific activity does not provide enough evidential support for a general thesis or a general principle about the nature of science. To see why, suppose that I study in great detail August Kekulé's discovery of the structure of the benzene molecule. I investigate historical documents as well as his own account of the discovery. Kekulé describes his discovery as follows:

> I was sitting, writing at my text-book; but the work did not progress; my thoughts were elsewhere. I turned my chair to the fire and dozed. Again the atoms were gamboling before my eyes. This time the smaller groups kept modestly in the background. My mental eye, rendered more acute by the repeated visions of the kind, could now distinguish larger structures of manifold conformation: long rows, sometimes more closely fitted together; all twining and twisting in snake-like motion. But look! What was that? One of the snakes had seized hold of its own tail, and the form whirled mockingly before my eyes. As if by a flash of lightning I awoke; and this time also I spent the rest of the night in working out the consequences of the hypothesis (quoted in Rodricks 1992, p. 7).

What I have, then, is a case study, that is, a particular, detailed description of a scientific activity. Now suppose that, based on a very detailed analysis of this particular act of scientific discovery by an individual scientist, I make the following argument that proceeds from this particular case study to a general conclusion:

> (P) August Kekulé discovered the ring shape of the benzene molecule while having a vision or a dream of a snake biting its own tail.
> Therefore,
> (C) Scientific discoveries occur when scientists have visions or dreams.

Now let us follow the decision procedure for the analysis and evaluation of arguments in natural language outlined in Chap. 1 (see Sect. 1.1) in order to find out if this is a good argument. Since we already have the premise and the conclusion in canonical (or standard) form, we simply need to figure out what type of argument it is supposed to be: deductive or inductive. Is the premise (P) of this argument supposed to provide conclusive or probable support for the conclusion (C)? If premise (P) were supposed to provide logically conclusive support for the conclusion, then it would fail to do so. After all, the fact that Kekulé had a eureka moment while having a dream or a vision does not necessarily mean that scientists in general have eureka moments while they have dreams or visions. Kekulé's particular act of discovery may simply be atypical. So, if the argument from the Kekulé case study were supposed to be a deductive argument, it would be an invalid argument.

Even if it fails as a deductive argument, the argument from the Kekulé case study could still be a good inductive argument. Is the premise of this argument supposed to provide probable support for the conclusion? In other words, if premise (P) were true, would it make the conclusion more likely to be true as well? Well, not really! Again, Kekulé's particular act of discovery may simply be an outlier. For all we know, Kekulé's eureka moment could have been a unique experience as far as scientific discoveries are concerned. The case study itself gives us no reason to believe that Kekulé's experience is typical of scientific discoveries in general. If Kekulé's experience is an outlier, and we have no reason to think that it is not, then the Kekulé case study does not make it more likely that scientific discoveries in general follow the same pattern as Kekulé's discovery of the structure of the benzene molecule. In other words, even if premise (P) were true, as the case study allegedly shows, it would fail to make the conclusion more probable or likely to be true. So, if the argument from the Kekulé case study were supposed to be an inductive argument, it would be a weak argument.

Of course, we could easily turn the argument from the Kekulé case study into a valid argument by adding premises. For example, we could turn it into a valid argument by adding a premise as follows:

(P1) August Kekulé discovered the ring shape of the benzene molecule after having a vision or a dream of a snake biting its own tail.

(P2) All scientists make scientific discoveries in *exactly* the same way Kekulé made his discovery.

Therefore,

(C) Scientific discoveries occur when scientists have visions or dreams.

With the additional premise (P2), the argument from the Kekulé case study is now a valid argument. For the premises, namely, (P1) and (P2), if true, do provide logically conclusive support for the conclusion. The problem, however, is that we have no reasons to believe that (P2) is true, and the case study provides no such reasons. After all, the case study is about Kekulé in particular; it is not about what all scientists generally do. So, while we can turn the argument from the Kekulé case study into a valid argument quite easily by adding premises, those additional premises would not be supported by the case study itself, and so the revised argument, although valid, could not be said to be a sound argument.

Likewise, we could easily turn the argument from the Kekulé case study into a strong argument by adding premises. For example, we could turn it into a strong argument by adding a premise as follows:

(P1) August Kekulé discovered the ring shape of the benzene molecule after having a vision or a dream of a snake biting its own tail.

(P2) Scientists make scientific discoveries in *pretty much* the same way Kekulé made his discovery.

Therefore,

(C) Scientific discoveries occur when scientists have visions or dreams.

With the additional premise (P2), the argument from the Kekulé case study is now a strong argument. For the premises, namely, (P1) and (P2), if true, provide probable support for the conclusion. The problem, however, is that we have no reasons to believe that (P2) is true, and the case study provides no such reasons. After all, the case study is about Kekulé in particular; it is not about what other scientists typically do. So, while we can turn the argument from the Kekulé case study into a strong argument quite easily by adding premises, those additional premises would not be supported by the case study itself, and so the revised argument, although strong, could not be said to be a cogent argument.

Our evaluation of the argument from the case study of Kekulé's discovery of the structure of the benzene molecule yields the verdict that it is a fallacious argument. A fallacious argument is an argument whose premises fail to provide either conclusive or probable support for its conclusion. In that sense, invalid and weak arguments are fallacious arguments. Now let us go back to the question we asked prior to considering the Kekulé case study. The question was this: Can a particular, detailed description of a scientific activity (that is, a case study) provide evidential support (either logically conclusive, as in deductive arguments, or probable, as in inductive arguments) for a thesis or principle in the scientific realism/antirealism debate in philosophy of science? Or, as Bishop and Trout (2002, p. S204) ask, "How much support does a single case study (or even a number of case studies) provide a general principle about the nature of science?" The answer is that a particular case or a single case study does not provide enough evidential support (either logically conclusive, as in deductive arguments, or probable, as in inductive arguments) for any general theses or principles about the nature of science. As far as the Kekulé case study is concerned, a particular act of scientific discovery by one scientist does not support any conclusions about acts of scientific discoveries in general. The same can be said about any case study because, case studies from the history of science are, by definition, "*particular*, detailed descriptions of scientific activity" (Currie 2015, p. 553), whereas the conclusions that philosophers of science are typically interested in, especially as far as the contemporary scientific realism/antirealism debate is concerned, are *general* conclusions or "general principle[s] about the nature of science" (Bishop and Trout 2002, p. S204). As Joseph Pitt (2001, p. 373) puts it, "if one starts with a case study, it is not clear where to go from there—for it is unreasonable to generalize from one case or even two or three."

Indeed, most textbooks of logic, reasoning, and argumentation classify generalizing "from one case or even two or three" as fallacious reasoning. For example, according to Patrick Hurley (2006, p. 131), "Hasty generalization is a fallacy that affects inductive reasoning. [...] The fallacy occurs when there is a reasonable likelihood that the sample [or case] is not representative of the group. Such a likelihood may arise if the sample is either too small or not randomly selected."[6] Likewise, according to Merrilee Salmon (2013, p. 151), to draw general conclusions from "a single vivid case" is a mistake in reasoning, a "fallacy of misleading vividness." Accordingly, if the particular case of Kekulé's discovery of the structure of the benzene molecule is not representative of scientific discoveries in general, then generalizing from this case would be an instance of hasty generalization. Since a sample that contains a particular case, such as that of Kekulé's discovery of the structure of the benzene molecule, is too small (after all, it contains only one case) and not randomly selected, we have reasons to suspect that it is not representative of scientific discoveries in general. Clearly, if the Kekulé' case study is not representative of scientific discoveries in general, then it cannot support any general theses or principles about the nature of scientific discoveries.[7]

In the context of the scientific realism/antirealism debate in contemporary philosophy of science, perhaps the most widely discussed case study from the history of science is probably the case of phlogiston. Both scientific realists and antirealists discuss this case study extensively. Antirealists have used it to argue against scientific realism (of some variety or another), whereas scientific realists have used it to argue for scientific realism (of some variety or another). For example, antirealists like Brad Wray (2018, p. 70) claim that "the concept of 'phlogiston' has no place in modern chemistry." For antirealists, phlogiston is an example of a theoretical posit that was postulated to explain a natural phenomenon, namely, combustion, but later turned out not to exist at all. As Wray (2018, p. 70) argues, "'phlogiston' was, in one sense, replaced by 'oxygen', though the types of substances designated by these terms have very different properties." For this reason, Wray (2018, p 70) thinks that "Only the most Whig historians of science could claim that oxygen is just phlogiston by another name."

As one might expect, scientific realists disagree. Indeed, some realists have argued exactly what Wray says would amount to Whiggish history of science, namely, that eighteenth century chemists were using phlogiston-based terminology, such as "dephlogisticated air," to talk about oxygen. As Miriam Solomon (2001, p. 37) explains:

---

[6] Ludwig Fahrbach (2011), Seungbae Park (2011), and I (Mizrahi 2013) have made this criticism against the inductive version of the antirealist argument commonly known as the "pessimistic induction" (that is, that it is an inductive generalization from a small and insufficiently diverse sample). This argument is analyzed and evaluated in Chap. 5, Sect. 5.1.

[7] For more on the problems with the method of using case studies as evidence for philosophical theses about science, see Mizrahi (2018, 2020).

"Phlogiston" will turn out to have reference at least sometimes because "dephlogisticated air" was successfully--*albeit unwittingly*--used in a particular set of experimental contexts to refer to oxygen (Kitcher 1993, p. 100). That is, "phlogiston" sometimes refers even though there is no such thing as Priestley's phlogiston, and even though no phlogiston theorist at the time could say how the term refers. Phlogiston theorists were actually often successfully referring to oxygen when they used the phrase "dephlogisticated air," although they did not know it (emphasis in original).

Who has the historical facts exactly right--scientific realists or antirealists--is difficult to tell. For our purposes, however, the important point is not historical but rather *logical*. That is to say, even if antirealists are right in thinking that eighteenth century chemists posited phlogiston to explain combustion, but phlogiston does not exist, no general conclusions about the nature of science would follow from that fact (even if it is a historical fact). In other words, the following antirealist argument is neither a valid nor a strong argument:

(P) Eighteenth century chemists posited the existence of a substance called "phlogiston" to explain combustion, but phlogiston does not really exist.
Therefore,

(C) All (or most of) the theoretical posits of science do not really exist.

As we have seen, a sample of one is too small to draw any general conclusions about the nature of science. As we have also seen, we could easily make this antirealist argument valid (or strong) by adding premises. For example:

(P1) Eighteenth century chemists posited the existence of a substance called "phlogiston" to explain combustion, but phlogiston does not really exist.
(P2) All (or most of) the theoretical posits of science are like phlogiston.

Therefore,

(C) All (or most of) the theoretical posits of science do not really exist.

Now this antirealist argument can be said to be valid (or strong), but it cannot be said to be sound (or cogent). This is because the premise we have added, namely, (P2), is an unsubstantiated premise, and the phlogiston case study gives us no reasons to believe that (P2) is true. For the phlogiston case study is about phlogiston in particular, not about theoretical posits in general.

Likewise, even if scientific realists are right in thinking that eighteenth century chemists used the phrase "dephlogisticated air" to talk about oxygen in combustion, no general conclusions about the nature of science would follow from that fact (even if it is a historical fact). In other words, the following realist argument is neither a valid nor a strong argument:

(P) Although they called it "dephlogisticated air" rather than "oxygen," eighteenth chemists posited the existence of a substance that really does play a role in combustion.
Therefore,

(C) All (or most of) the theoretical posits of science really do exist.

As we have seen, a sample of one is too small to draw any general conclusions about the nature of science. As we have also seen, we could easily make this realist argument valid (or strong) by adding premises. For example:

(P1) Although they called it "dephlogisticated air" rather than "oxygen," eighteenth chemists posited the existence of a substance that really does play a role in combustion.
(P2) All (or most of) the theoretical posits of science are like dephlogisticated air.

Therefore,

(C) All (or most of) the theoretical posits of science really do exist.

Now this realist argument can be said to be valid (or strong), but it cannot be said to be sound (or cogent). This is because the premise we have added, namely, (P2), is an unsubstantiated premise, and the phlogiston case study gives us no reasons to believe that (P2) is true. For the phlogiston case study is about phlogiston in particular, not about theoretical posits in general. So, once again, "if one starts with a case study, it is not clear where to go from there—for it is unreasonable to generalize from one case or even two or three" (Pitt 2001, p. 373).

For these reasons, rather than give readers detailed descriptions of a few case studies from the history of science, which are probably a recipe for fallacious arguments, this book takes an argumentation approach to the scientific realism/antirealism debate in contemporary philosophy of science. In Chap. 4, I will analyze and evaluate what I take to be key arguments for scientific realism using the argumentation approach described above. In Chap. 5, I will analyze and evaluate what I take to be key arguments against scientific realism (or for antirealism) using the argumentation approach described above. Before we get to the analyses of what I take to be key arguments in the debate, however, I will first survey what I take to be key positions in debate in Chap. 3. Some of these positions are realist positions, broadly speaking, whereas others are antirealist positions, broadly speaking. Finally, in Chap. 6, I will discuss my own brand of scientific realism, namely, Relative Realism. I take Relative Realism to be a middle ground position between scientific realism and antirealism. I have proposed this view for the first time in a paper published in *International Studies in the Philosophy of Science* in 2013. But I will develop this position in much more detail, as well as advance novel arguments for it, in Chap. 6 of this book.

## 2.3  Summary

Scientific realism has three dimensions: a metaphysical dimension, a semantic dimension, and an epistemic dimension. The metaphysical thesis (or stance or dimension) of scientific realism is supposed to capture the idea that there are things out there in the world for scientists to discover and that those things out there in the world are independent of the human minds that study them. Antirealists tend to reject this metaphysical stance with respect to the unobservable entities, processes, and events that figure in our best scientific theories. The semantic thesis (or stance or dimension) of scientific realism is supposed to capture the idea that scientific theories are to be taken literally, which means that they can be either true or false. Some antirealists reject this semantic stance. The epistemic thesis (or stance or dimension) of scientific realism is supposed to capture the idea that there are good reasons to believe that our best scientific theories, in particular, those that are empirically successful, are approximately true. Antirealists tend to reject this epistemic stance with respect to the theoretical claims about unobservable entities, processes, and events that figure in our best scientific theories. Since scientific realists and antirealists aim to argue for and/or against general theses about the nature of science, a case study (or even a few case studies) cannot provide evidential support (either logically conclusive, as in deductive arguments, or probable, as in inductive arguments) for the sort of theses that scientific realists and antirealists argue for and/or against in the scientific realism/antirealism debate in contemporary philosophy of science.

## Glossary

**Antirealism** An agnostic or skeptical attitude toward the theoretical posits (that is, unobservables) of scientific theories. Antirealism comes in different varieties, such as Constructive Empiricism (see Chap. 3, Sect. 3.3) and Instrumentalism (see Chap. 3, Sect. 3.2).

**Approximate truth** Closeness to the truth or truthlikeness. To say that a theory is approximately true is to say that it is close to the truth. According to some scientific realists, approximate truth is the aim of science. (See Sect. 2.1).

**Case study** A particular, detailed description of a scientific activity, a scientific practice, or an episode from the history of science. (See Sect. 2.2.)

**Direct observation** Observation with the naked eye, without the use of scientific instruments, such as microscopes and telescopes, as opposed to instrument-aided observation. (See Chap. 3, Sect. 3.3.)

**The epistemic dimension (or stance) of scientific realism** The thesis that our best scientific theories, in particular, those that are empirically successful, are approximately true. (See Sect. 2.1.)

**Fallacious argument** An argument whose premises fail to provide either conclusive or probable support for its conclusion (see also *invalid argument* and *weak argument*). (See Sect. 2.2.)

**Hasty generalization** A fallacious inductive argument from a sample that is not representative of the general population that is the subject of the conclusion of the argument (because the sample is too small or cherry-picked rather than randomly selected). (See Sect. 2.2.)

**Instrument-aided observation** Observation by means of scientific instruments, such as microscopes and telescopes, as opposed to direct or naked-eye observation. (See Chap. 3, Sect. 3.3.)

**Invalid argument** A deductive argument in which the premises purport but fail to provide logically conclusive support for the conclusion. (See Chap. 1, Sect. 1.1.)

**The metaphysical dimension (or stance) of scientific realism** The thesis that there are things out there in the world for scientists to discover and that those things out there in the world are independent of the human minds that study them. (See Sect. 2.1.)

**Scientific realism** An epistemically positive attitude toward those aspects of scientific theories that are worthy of belief. Scientific realism comes in different varieties, such as Explanationist Realism (see Chap. 3, Sect. 3.1), Entity Realism (see Chap. 3, Sect. 3.4), Structural Realism (see Chap. 3, Sect. 3.5), and Relative Realism (see Chap. 6, Sect. 6.1).

**The semantic dimension (or stance) of scientific realism** The thesis that scientific theories are to be taken literally, which means that they can be either true or false. (See Sect. 2.1.)

**Weak argument** A non-deductive (or inductive) argument in which the premises purport but fail to provide probable support for the conclusion. (See Chap. 1, Sect. 1.1.)

## References and Further Readings

Aronson, J. L., Harré, R., & Way, E. C. (1994). *Realism rescued: How scientific progress is possible*. London: Duckworth.

Bishop, M. A., & Trout, J. D. (2002). 50 years of successful predictive modeling should be enough: Lessons for philosophy of science. *Philosophy of Science, 69*(September 2002), S197–S208.

Boyd, R. (1990). Realism, approximate truth and philosophical method. In W. C. Savage (Ed.), *Scientific theories* (pp. 355–391). Minneapolis: University of Minnesota Press.

Chakravartty, A. (2007). *A metaphysics for scientific realism: Knowing the unobservable*. New York: Cambridge University Press.

Chakravartty, A. (2010). Truth and representation in science: Two inspirations from art. In R. Frigg & M. Hunter (Eds.), *Beyond mimesis and convention: Representation in art and science* (pp. 33–50). Dordrecht: Springer.

Chakravartty, A. (2017). Scientific realism. In E. N. Zalta (Ed.), *The Stanford encyclopedia of philosophy* (Summer 2017 ed.). https://plato.stanford.edu/archives/sum2017/entries/scientific-realism/

Currie, A. (2015). Philosophy of science and the curse of the case study. In C. Daly (Ed.), *The Palgrave handbook of philosophical methods* (pp. 553–572). London: Palgrave Macmillan.

Fahrbach, L. (2011). How the growth of science ends theory change. *Synthese, 180*(2), 139–155.

Hurley, P. J. (2006). *A concise introduction to logic* (9th ed.). Belmont: Wadsworth.

Ivanova, M. (2013). Did Perrin's experiments convert Poincaré to scientific realism? *HOPOS: The Journal of the International Society for the History of Philosophy of Science, 3*(1), 1–19.

Kitcher, P. (1989). Explanatory unification and the causal structure of the word. In P. Kitcher & W. Salmon (Eds.), *Scientific explanation* (pp. 410–505). Minneapolis: University of Minnesota Press.

Kitcher, P. (1993). *The advancement of science: Science without legend, objectivity without illusions*. New York: Oxford University Press.

Kitcher, P. (2002). Scientific knowledge. In P. K. Moser (Ed.), *The Oxford handbook of epistemology* (pp. 385–407). New York: Oxford University Press.

Laudan, L. (1981). A confutation of convergent realism. *Philosophy of Science, 48*(1), 19–49.

Laymon, R. (1984). The path from data to theory. In J. Leplin (Ed.), *Scientific realism* (pp. 108–123). Berkeley: University of California Press.

Leplin, J. (1981). Truth and scientific progress. *Studies in History and Philosophy of Science Part A, 12*(4), 269–291.

Miller, D. (1974). Popper's qualitative theory of verisimilitude. *The British Journal for the Philosophy of Science, 25*(2), 166–177.

Mizrahi, M. (2013). The pessimistic induction: A bad argument gone too far. *Synthese, 190*(15), 3209–3226.

Mizrahi, M. (2018). Introduction. In M. Mizrahi (Ed.), *The Kuhnian image of science: Time for a decisive transformation?* (pp. 1–22). London: Rowman & Littlefield.

Mizrahi, M. (2020). The case study method in philosophy of science: An empirical study. *Perspectives on Science, 28*(1), 63–88.

Needham, P. (2018). Scientific realism and chemistry. In J. Saatsi (Ed.), *The Routledge handbook of scientific realism* (pp. 345–356). New York: Routledge.

Niiniluoto, I. (1998). Verisimilitude: The third period. *The British Journal for the Philosophy of Science, 49*(1), 1–29.

Park, S. (2011). A confutation of the pessimistic induction. *Journal for General Philosophy of Science, 42*(1), 75–84.

Pew Research Center. (2015). *Public and scientists' views on science and society.* Pew Research Center, January 29, 2015. https://www.pewinternet.org/wp-content/uploads/sites/9/2015/01/PI_ScienceandSociety_Report_012915.pdf. Accessed on 14 Aug 2019.

Pitt, J. C. (2001). The dilemma of case studies: Toward a Heraclitian philosophy of science. *Perspectives on Science, 9*(4), 373–382.

Popper, K. R. (1972). *Conjectures and refutations: The growth of scientific knowledge* (4th ed.). London: Routledge.

Post, H. R. (1971). Correspondence, invariance and heuristics: In praise of conservative induction. *Studies in History and Philosophy of Science Part A, 2*(3), 213–255.

Psillos, S. (1999). *Scientific realism: How science tracks truth.* London: Routledge.

Psillos, S. (2018). Realism and theory change in science. In E. N. Zalta (Ed.), *The Stanford encyclopedia of philosophy* (Summer 2018 ed.). https://plato.stanford.edu/archives/sum2018/entries/realism-theory-change

Psillos, S., & Ruttkamp-Bloem, E. (2017). Scientific realism: Quo vadis? Introduction: New thinking about scientific realism. *Synthese, 194*(9), 3187–3201.

Rodricks, J. V. (1992). *Calculated risks: Understanding the toxicity and human health risks of chemicals in our environment.* New York: Cambridge University Press.

Rowbottom, D. P. (2019). Scientific realism: What it is, the contemporary debate, and new directions. *Synthese, 196*(2), 451–484.

Salmon, M. H. (2013). *Introduction to logic and critical thinking* (6th ed.). Boston: Wadsworth.

Sankey, H. (2008). *Scientific realism and the rationality of science.* Hampshire: Ashgate.

Solomon, M. (2001). *Social empiricism.* Cambridge, MA: The MIT Press.

Tichý, P. (1974). On Popper's definitions of verisimilitude. *The British Journal for the Philosophy of Science, 25*(2), 155–160.

van Fraassen, B. C. (1980). *The scientific image.* Oxford: Oxford University Press.

Wolff, J. (2019). Naturalistic quietism or scientific realism? *Synthese, 196*(2), 485–498.

Wray, B. K. (2018). *Resisting scientific realism.* Cambridge: Cambridge University Press.

# Chapter 3
# Key Positions in the Contemporary Scientific Realism/Antirealism Debate

**Abstract**   In this chapter, I survey key positions in the scientific realism/antirealism debate in contemporary philosophy of science. The first is a selective realist position, which is known as Explanationist Realism, according to which we should believe only in the indispensable parts of our best scientific theories. Those parts that are considered indispensable, the so-called "working posits" (Kitcher, The advancement of science: Science without legend, objectivity without illusions. Oxford University Press, New York, 1993), are the ones that are responsible for, or best explain, the predictive success of our best scientific theories (Psillos, Scientific realism: how science tracks truth. Routledge, London, 1999). The second is an antirealist position with a long history in philosophy of science (see, for example, Duhem, The aim and structure of physical theory. Princeton University Press, Princeton, NJ. Translated from the French by Philip P. Wiener, 1954/1982), known as Instrumentalism, according to which scientific theories are mere instruments or tools of prediction (Rowbottom, The instrument of science: scientific anti-realism revitalised. Routledge, London, 2019). The third is an influential antirealist position, which is known as Constructive Empiricism and is due to Bas van Fraassen (The scientific image. Oxford University Press, New York, 1980), according to which science aims at empirical adequacy, not truth or approximate truth. The fourth is another influential realist position, which is known as Entity Realism and is due to Ian Hacking (Representing and intervening: introductory topics in the philosophy of natural science. Cambridge University Press, New York, 1983), according to which one is justified in taking a realist stance with respect to entities that can be manipulated and that facilitate interventions in nature (Sankey, Howard, Scientific realism and the rationality of science. Ashgate, Hampshire, 2008). The fifth is another selective realist position, which is known as Structural Realism, according to which we should be realists, not about theoretical entities or processes, but rather about structures (Worrall, Dialectica 43(1–2):99–124, 1989; Ladyman, Stud. Hist. Philos. Sci. A 29(3):409–424, 1998; French, The structure of the world: metaphysics and representation. Oxford University Press, Oxford, 2014).

**Keywords**   Belief/acceptance distinction · Constructive empiricism · Empirical adequacy · Empirical success · Entity realism · Epistemic structural Realism (ESR) · Explanationist Realism (also known as deployment realism) · Instrumentalism · Observation/detection distinction · Ontic structural Realism (OSR) · Predictive success · Structural realism · Working posits/idle parts distinction

© Springer Nature Switzerland AG 2020                                                     35
M. Mizrahi, *The Relativity of Theory*, Synthese Library 431,
https://doi.org/10.1007/978-3-030-58047-6_3

In this chapter, I survey key positions in the scientific realism/antirealism debate in contemporary philosophy of science. The first is a selective realist position, which is known as Explanationist Realism, according to which we should believe only in the indispensable parts of our best scientific theories. Those parts that are considered indispensable, the so-called "working posits" (Kitcher 1993), are the ones that are responsible for, or best explain, the predictive success of our best scientific theories (Psillos 1999). The second is an antirealist position with a long history in philosophy of science (see, for example, Duhem 1954), known as Instrumentalism, according to which scientific theories are mere instruments or tools of prediction (Rowbottom 2019). The third is an influential antirealist position, which is known as Constructive Empiricism and is due to Bas van Fraassen (1980), according to which science aims at empirical adequacy, not truth or approximate truth. The fourth is another influential realist position, which is known as Entity Realism and is due to Ian Hacking (1983), according to which one is justified in taking a realist stance with respect to entities that can be manipulated and that facilitate interventions in nature (Sankey 2008). The fifth is another selective realist position, which is known as Structural Realism, according to which we should be realists, not about theoretical entities or processes, but rather about structures (Worrall 1989; Ladyman 1998; French 2014).

## 3.1   Explanationist Realism

Both scientific realists and antirealists agree that science is successful. When philosophers of science talk about "the success of science," they usually mean *empirical success*, which includes explanatory and predictive success. That is to say, the best scientific theories are those that explain natural phenomena that would otherwise be mysterious to us and that make predictions that are borne out by observation and experimentation. More specifically, in the context of the scientific realism/antirealism debate in contemporary philosophy of science, the sort of success that is particularly important for both realists and antirealists is predictive success. That is to say, there are scientific theories that make predictions that are borne out by observational and experimental testing. For example, according to Albert Einstein's theory of General Relativity, the fabric of space-time can be distorted or "warped" by the presence of massive objects, such as stars and black holes. This means that, when massive objects, such as neutron stars, move rapidly through space-time, they cause ripples in space-time, in much the same way that stones dropped into a pond produce a ripple effect. Instead of waves in water, however, moving masses like neutron stars produce gravitational waves that ripple across space-time. In other words, Einstein's theory of General Relativity predicts the existence of gravitational waves. Einstein proposed his theory of General Relativity in 1915. Subsequently, the existence of gravitational waves was confirmed indirectly by the discovery of a binary pulsar in 1974 (for which Russell A. Hulse and Joseph H. Taylor, Jr. were awarded the Nobel Prize in Physics in 1993) and directly by the Laser Interferometer

Gravitational-Wave Observatory (LIGO) in 2015 (Abbott et al. 2016). This is quite astonishing: a prediction made by a theory that was proposed at the beginning of the twentieth century is confirmed by experimentation roughly a century later. What could explain this astonishing predictive success of scientific theories like that of Einstein's theory of General Relativity?

Explanationist Realism is the view that realist commitments are warranted with respect to the theoretical posits that are responsible for--or can best explain--the predictive success of our best scientific theories, such as the successful prediction of Einstein's theory of General Relativity. This realist view is based on Stathis Psillos' distinction between the theoretical posits of a scientific theory that are responsible for its predictive success (that is, for the fact that the theory makes predictions that are borne out by observation and experimentation) and those that are not responsible for the predictive success of the theory. As Psillos (1999, pp. 112–113) puts it:

> It is precisely *those theoretical constituents which scientists themselves believe to contribute to the successes of their theories* (and hence to be supported by the evidence) that tend to get retained in theory change. Whereas, the constituents that do not 'carry-over' tend to be those that scientists themselves considered too speculative and unsupported to be taken seriously (emphasis added).[1]

Accordingly, realists of the explanationist stripe argue that we should follow the lead of scientists in believing in the existence of those theoretical posits that contribute to or are responsible for the predictive success of a scientific theory. For example, realists of the explanationist stripe would argue that gravitational waves are one of those theoretical constituents that physicists believe to contribute to the success of Einstein's theory of General Relativity, and are supported by the evidence from the Laser Interferometer Gravitational-Wave Observatory (LIGO) experiments, which uses interferometers to detect the kind of interference patterns that are associated with waves.

Some parties to the scientific realism/antirealism debate in contemporary philosophy of science refer to Explanationist Realism as "Deployment Realism." For example, Timothy Lyons (2016, p. 95) writes:

> According to deployment realism, we can be justified in believing the following meta-hypothesis: *those theoretical constituents that were genuinely deployed in the derivation of novel predictive success are at least approximately true* (emphasis added).[2]

Lyons (2016, p. 95) goes on to point out that the argument for this realist meta-hypothesis is that it would be a miracle if the scientific theories that feature these theoretical constituents that are genuinely deployed in the derivation of novel

---

[1] Philip Kitcher (1993, p. 149) draws a somewhat similar distinction between "*working posits* (the putative referents of terms that occur in problem-solving schemata) and *presuppositional posits* (those entities that apparently have to exist if the instances of the schemata are to be true)" (emphasis in original).

[2] When philosophers of science talk about "novel predictions," they typically mean a prediction that was not known to be true (or was expected to be true or false) at the time the theory was constructed.

predictions were not even approximately true. For a detailed analysis and evaluation of this argument for scientific realism, commonly known as the "no miracles" argument, see Chap. 4, Sect. 4.1. For now, the important point is that scientific realists do not have to be realists about all the theoretical posits of scientific theories. Instead, realists can be more *selective* about the entities that deserve their realist commitments.

Given this distinction between the "working posits" (that is, those theoretical posits that are responsible for the predictive success of a scientific theory) and "idle parts" (that is, those theoretical posits that do not play any part in explaining the predictive success of a scientific theory) of a scientific theory, realists of the explanationist stripe go on to argue that "it is enough to show that the theoretical laws and mechanisms which generated the successes of past theories have been retained in our current scientific image" (Psillos 1999, p. 108). This allows realists of the explanationist stripe to be optimistic about the progress of science even if many theoretical posits of past scientific theories have been abandoned, as the history of science allegedly shows. (On the pessimistic argument from the history of science, see Chap. 5, Sect. 5.1). The key argument for Explanationist Realism is known as the "miracle" argument or the "no miracles" argument. This argument is analyzed and evaluated in Chap. 4, Sect. 4.1.

## 3.2   Instrumentalism

As mentioned in Sect. 3.1, there are scientific theories that make predictions that are borne out by experimentation and observation. Some antirealists think that we should focus on the predictive success of scientific theories and stop worrying about whether those theories are likely true or approximately true. In fact, if science is a cognitive tool, then scientific theories should be understood as means to an end. As Darrell Rowbottom (2019, p. 1) puts it, "science is primarily, and should primarily be, an instrument for furthering our practical ends." This is the core tenet of an antirealist position known as Instrumentalism.

Instrumentalism is the view that scientific theories are instruments for attaining practical goals, such as predicting the occurrence of natural phenomena. For example, Pierre Duhem (1954/1982) expresses an instrumentalist view of scientific theories when he says that "A Law of Physics Is, Properly Speaking, neither True nor False" (p. 168) and that "propositions introduced by a theory [...] are neither true nor false; they are only convenient or inconvenient" (p. 334).[3] Understood as convenient or inconvenient instruments, then, rather than as attempts to describe the underlying nature of reality, scientific theories should not be taken literally, as either true or false, but rather instrumentally, as useful or not useful. In that respect,

---

[3] Milena Ivanova (2013) provides a useful discussion of Duhem and realism about atoms and the atomic theory.

instrumentalists reject the semantic stance (or thesis or dimension) of scientific realism (see Chap. 2, Sect. 2.1).

Accordingly, instrumentalists recommend that we refrain from interpreting theoretical statements literally. For example, we should not take the theoretical statement "SARS-CoV-2 is the virus that causes COVID-19" as a literal statement about an unobservable entity called "the novel coronavirus" or "SARS-CoV-2" that causes infected patients to have symptoms of an infectious disease, such as fever and shortness of breath. Instead, we should treat this theoretical statement as an instrument for explaining and predicting the appearance of symptoms in patients. If it is useful in explaining and predicting the onset of those symptoms, such as fever and shortness of breath, then we are entitled to use this instrument. What matters for instrumentalists is the usefulness of scientific theories as tools for accomplishing practical goals, such as explaining and predicting the occurrence of natural phenomena, not whether they are likely true or approximately true.

Likewise, the theoretical statement "HIV infection is caused by the human immunodeficiency virus" should be taken as neither true nor false, but rather as a convenient or useful tool for explaining and predicting the spread of the chronic condition called Acquired Immunodeficiency Syndrome (AIDS). For instrumentalists, we should not worry about whether pathogenic microorganisms and viruses, such as the Human Immunodeficiency Virus (HIV), are real or not. As long as they serve us in explaining and predicting the spread of infectious diseases, we are entitled to use them as such, that is, as instruments or tools, provided that we are not tempted to say that theoretical statements about such entities are literally true (or literally false). In that respect, Instrumentalism belongs to "an empiricist philosophical tradition" (Rowbottom 2019, p. 1), which gives epistemic priority to direct observation. Since pathogenic microorganisms, such as bacteria and viruses, are unobservable, that is, they cannot be directly observed with the naked eye, we should not commit ourselves to their existence by believing that any theoretical statements about them are literally true (or literally false). Instrumentalism is supported by several antirealist arguments, which are analyzed and evaluated in Chap. 5.

## 3.3  Constructive Empiricism

Like instrumentalism, another antirealist position that gives epistemic priority to direct observation is known as Constructive Empiricism. As Bas van Fraassen (1980, p. 12) puts it, Constructive Empiricism is the view that "Science aims to give us theories that are empirically adequate; and acceptance of a theory involves as belief only that it is empirically adequate." For van Fraassen (1980, p. 18), "to accept a theory is (for us) to believe that it is empirically adequate—that what the theory says about what is observable (by us) is true." For constructive empiricists, what counts as "observable" must be directly observable (by us) without the aid of any instruments. As van Fraassen (2001, p. 154) puts it:

> We can *detect* the presence of things and the occurrence of events by means of instruments. But in my book that does not generally count as *observation*. Observation is perception, and perception is something that is possible for us without instruments (emphasis added).

In other words, when scientists use instruments, such as electron microscopes and radio telescopes, they are not *observing* theoretical entities, such as viruses and cosmic radiation, according to constructive empiricists. Rather, scientists are merely *detecting* these things. For example, van Fraassen (2001, p. 158) argues that paramecia, a type of single-celled (or unicellular) microorganisms, can be detected by using microscopes, but they cannot be directly observed.

> The microscope's output can be sent into a scanner which transmits to a computer or projector – then we see the paramecia on the wall or the monitor. *We are having a different sort of experience* then, for we say after only a little urging that we are seeing an image. Nothing about the status of the microscope can follow, it seems to me, from concentration on one of these three experimental arrangements to the exclusion of the others (emphasis added).

If one is inclined to say that one sees an image of an object (for example, a paramecium) through the microscope, van Fraassen (2001, p. 160) argues, then one is no longer simply "gathering information" but rather "postulating" that there is a real object under the microscope. This emphasis on direct observation (as opposed to mere detection) leads constructive empiricists to recommend suspension of belief in any theoretical statements that would commit us to the existence of unobservable entities, processes, or events.

Constructive Empiricism, then, is the view that, when scientists construct theories about natural phenomena, they are not aiming for literally true descriptions of the underlying nature of reality. Rather, they are merely aiming to "save the phenomena," that is, to get the observable facts right. Here is how van Fraassen (1998, p. 213) contrasts Constructive Empiricism with scientific realism:

> Scientific realism and constructive empiricism are, as I understand them, not epistemologies but views of what science is. Both views characterize science as an activity with an aim--a point, a criterion of success--and construe (unqualified) acceptance of science as involving the belief that science meets that criterion. According to scientific realism the aim is truth (literally true theories about what things are like). *Constructive empiricism sees the aim as not truth but empirical adequacy* (emphasis added).

If Constructive Empiricism is the correct view of what science is, then scientists do not (or should not) care whether the theoretical entities they posit to explain observable phenomena are real or not. Rather, scientists only care (or should only care) that the observable predictions derived from those theoretical postulates are accurate.[4] In that respect, constructive empiricists deny the epistemic stance (or thesis or dimension) of scientific realism (see Chap. 2, Sect. 2.1).

---

[4] In Mizrahi (2014), I argue that Constructive Empiricism is ambiguous between a descriptive reading and a normative reading. On the descriptive interpretation, Constructive Empiricism is the view that scientists *do* aim for empirically adequate theories. On the normative interpretation, Constructive Empiricism is the view that scientists *should* aim for empirically adequate theories.

Unlike instrumentalists, for whom scientific theories should not be understood literally, constructive empiricists do take scientific theories literally, that is, as capable of being true or false. For constructive empiricists, then, scientific theories are not mere instruments, as they are for instrumentalists. Rather, scientific theories consist of statements that can be true or false. However, we should only believe what scientific theories say about observable phenomena, constructive empiricists argue, and withhold belief from what scientific theories say about unobservable entities, processes, and events. For constructive empiricists, to believe what scientific theories say about unobservables is to adopt "beliefs going beyond what science itself involves or requires for its pursuit" (van Fraassen 1998, p. 214). In other words, constructive empiricists argue that there is no need to believe in the unobservable entities, processes, and events of scientific theories in order to make sense of the scientific enterprise. For constructive empiricists, belief in what scientific theories say about observable phenomena, that is, *acceptance* (or the belief that a theory is empirically adequate), is all that is required both to do science and to make sense of science. The main "Positive Argument" for Constructive Empiricism is analyzed and evaluated in Chap. 5, Sect. 5.2.

## 3.4 Entity Realism

Although the scientific realism/antirealism debate in contemporary philosophy of science is typically construed as a debate concerning the epistemic status of scientific theories (see Chap. 2, Sect. 2.1), this debate can also be understood as a debate concerning the ontological status of the theoretical posits of science. That is to say, instead of asking whether our best scientific theories are likely true or approximately true, or whether we are justified in believing our best scientific theories, we could ask whether the theoretical posits of our best scientific theories are real. (Recall the metaphysical thesis, stance, or dimension discussed in Chap. 2, Sect. 2.1). As Bence Nanay (2019, p. 500) argues, the "debate about theories is very different from, and logically independent of, the debate about unobservable entities."

To illustrate, consider the recent reports that scientists have managed to take a picture of a black hole. In *The New York Times*, the headline of an article announcing the discovery read as follows: "Darkness Visible, Finally: Astronomers Capture First Ever Image of a Black Hole" (Overbye 2019). Now, black holes are theoretical entities insofar as they are unobservable entities whose existence is postulated by scientific theories, specifically, Albert Einstein's theory of General Relativity. According to Einstein's theory of General Relativity, a massive amount of matter can be packed into a tiny region of space such that space-time itself is deformed. This deformity in space-time is known as a black hole. Since the gravitational pull of such a massive amount of matter in a small and dense region of space is so strong, nothing can escape the gravitational pull of a black hole once it passes what is known as the event horizon of a black hole. This means that a black hole does not give off any light, which in turn means that a black hole cannot be observed directly, given that "Eyes are devices for extracting useful information from the light reflected or emitted from objects in the world around us" (Land and Nilsson 2012, p. 23). Now, on April 10, 2019, the scientific journal, *Nature*, reported that the "Event Horizon Telescope's global network of radio dishes has produced the first-ever direct image of a black hole and its event horizon" (Castelvecchi 2019). This is a bit misleading, however, because what can be seen in that picture is the light at the edge of a black hole, that is, before the light crosses the event horizon. The presence of the black hole itself is then inferred from the fact that there appears to be blackness in the middle of that "photon ring," as the astronomer Andrea Ghez calls it in the *Nature* article (Castelvecchi 2019, p. 285). Another astronomer, Heino Falcke, makes the same point when he says, "What you're looking at is a ring of fire created by the deformation of space-time. Light goes around, and looks like a circle" (Castelvecchi 2019, p. 284). This "ring of fire" itself, however, is not a black hole. For, as we have seen, a black hole is a region of space that has a massive amount of mass concentrated in it such that nearby objects cannot escape its gravitational pull. In other words, the picture is not a direct image of a black hole. Nor is it an instance of darkness made visible, as *The New York Times* headline asserts. Rather, it is an image of the *light* around what is supposed to be the event horizon of a black hole.

Accordingly, antirealists would insist that whether black holes exist or not is still an open question, the announcements in *The New York Times* (Overbye 2019) and in *Nature* about "the first-ever direct image of a black hole" (Castelvecchi 2019) notwithstanding. For instance, constructive empiricists might invoke the distinction between *observing* and *detecting* to argue that scientists have detected light around what they think is a black hole, but they have not observed a black hole directly (see Sect. 3.3). Therefore, black holes remain theoretical posits whose existence is inferred from theoretical assumptions rather than directly observed without the aid of scientific instruments. For constructive empiricists, this means that we should not believe that black holes really exist. Instead, we should simply *accept* Einstein's theory of General Relativity without committing ourselves to the existence of its theoretical posits, such as black holes, gravitational waves, and the like.

Entity Realism is the view that the theoretical entities, processes, and events (that is, unobservables) posited by our best scientific theories are real. As Howard Sankey (2008, p. 43) puts it, "Entity realism is the thesis that the unobservable theoretical entities of science are real." Bence Nanay (2019, p. 500) agrees with Sankey when he says that, "Realism about entities is the view that unobservable entities that scientific theories postulate really do exist. And, at least according to the proponents of Entity Realism, we can be realist about an unobservable entity a theory postulates without being realist about the theory that postulates it." One of the key arguments for this view is an argument from the manipulation of theoretical entities, which is based on Ian Hacking's (1983, p. 23) famous slogan for Entity Realism, "if you can spray them, they are real." That is to say, if we can do things with the theoretical entities posited by our scientific theories, then we have a good reason to believe that these theoretical entities are real. To use Hacking's (1983, p. 23) example, we can shoot electrons through a double-slit apparatus using an electron gun. Presumably, we would not be able to do that if electrons did not really exist. Therefore, the fact that we can do things with electrons, such as shooting electrons through a double-slit apparatus using an electron gun, provides a good reason to believe that electrons are real. Similarly, we can combine genetic material from various sources to create recombinant DNA using methods of genetic recombination such as molecular cloning. Presumably, we would not be able to do that if DNA molecules did not really exist. Therefore, the fact that we can do things with DNA, such as combining genetic material from various sources to create recombinant DNA, provides a good reason to believe that DNA molecules are real.

Of course, black holes cannot be manipulated by us. Indeed, as we have seen, antirealists would argue that we cannot even observe black holes directly, let alone manipulate them. Hacking was aware of this. According to Hacking (1989, p. 561), "A black hole is as *theoretical* an entity as could be. Moreover, it is in principle *unobservable*. [...] At best we can *interpret* various phenomena as being due to the existence of black holes" (emphasis added). If, as entity realists argue, we should believe that a theoretical entity is real only if we can manipulate that entity, but black holes cannot be manipulated by us, then it follows that we should not believe that there are such things as black holes. As Hacking (1989, p. 578) puts it, "When we use entities as *tools*, as *instruments of inquiry*, we are entitled to regard them as real. But we cannot do that with the objects of astrophysics" (emphasis added), such as black holes. Therefore, by entity realists' lights, it appears that we are not entitled to regard the theoretical entities of astronomy and astrophysics as real. For Dudley Shapere (1993), this consequence of Entity Realism, namely, that "astronomy is not a natural science at all" (Hacking 1989, p. 577) because its theoretical entities, such as black holes, cannot be manipulated by us, was an absurd consequence of Entity Realism. Shapere interprets Hacking as arguing that "Astronomy cannot interfere with its objects" (Shapere 1993, p. 135), and so it follows that "astronomy is not a natural science at all" (Hacking 1989, p. 577). Shapere (1993, p. 135) argues that "astronomy is a science" and that Hacking is wrong about the "role of experiment in science." On the one hand, for Hacking, astronomy is a natural science only if it is an experimental science (that is, its theoretical entities can be used as tools or instruments of inquiry by us). Since astronomy is not an experimental science (that is, its theoretical entities, such as black holes, cannot be used as tools or instruments of inquiry by us), according to Hacking, it follows that "astronomy is not a natural science at all" (Hacking 1989, p. 577). On the other hand, for Shapere, astronomy is a natural science even if it is not an experimental science. Since "passively observing" can advance scientific knowledge just as much as "interfering actively" can, according to Shapere, it follows that "astronomy is a science" (Shapere 1993, p. 135). As Hilary Putnam would put it, this is a case in which "one philosopher's *modus ponens* is another philosopher's modus *tollens*" (Putnam 1994, p. 280). Aside from Hacking's argument from the manipulation of entities, another key argument for Entity Realism is an argument from corroboration. This argument is analyzed and evaluated in Chap. 4, Sect. 4.4.

## 3.5  Structural Realism

As we discussed in Chap. 2, the scientific realism/antirealism debate in contemporary philosophy of science has taken what might be called a "selectivist turn." That is to say, contemporary scientific realists are more *selective* about the aspects of science they are willing to be realists about than previous scientific realists were. We have already discussed two selective realist positions: Explanationist Realism (see Sect. 3.1) and Entity Realism (see Sect. 3.4). A third selective realist position in the contemporary scientific realism/antirealism debate is known as Structural Realism. For structural realists, the parts of our best scientific theories that warrant a realist commitment are unobservable *structures*, such as those represented by relations captured in the equations of our best scientific theories, rather than unobservable entities, processes, or events. On this view, we have no good reasons to believe that, say, Isaac Newton discovered real forces in nature, such as the force of gravity. But we do have good reasons to believe that "what Newton really discovered are the relationships between phenomena expressed in the mathematical equations of his theory [for example, his law of gravitation: $F_g = G\dfrac{M1\,M2}{r^2}$ ]"[5] (Worrall 1989, p. 122). For structural realists, "the relations of physical objects have all sorts of knowable properties, [but] the physical objects themselves remain unknown in their *intrinsic nature*" (Russell 1912, pp. 32–34; emphasis added).

The general motivation for this view comes from the role that mathematics plays in science. As Steven French (2014, p. 192) puts it:

> With the mathematization of science it is natural to extend this thesis [that is, the thesis that mathematics describes the structure of a domain] to scientific knowledge and then the latter too comes to be conceived as *structural knowledge* (emphasis added).[6]

Accordingly, structural realists argue that our best scientific theories do give us theoretical knowledge, but that theoretical knowledge is knowledge about *structures*, not entities, processes, or events. After all, if "the universe [...] is written in the language of mathematics, and its characters are triangles, circles, and other geometric figures without which it is humanly impossible to understand a single word of it," as Galileo (1623/1957, pp. 237–238) has argued, then one might think that scientists capture the mathematical structure of the universe in the mathematical equations of their scientific theories. It is in this sense that science can be said to give us structural knowledge of the universe.

More specifically, Structural Realism can be construed as either an epistemological position or an ontological position. Those who construe Structural Realism as an epistemological position subscribe to the view that our best scientific theories give us knowledge about the unobservable structure of the world. This view is known as

---

[5] Where $G$ is the gravitational constant, $M_1$ is the first mass, $M_2$ is the second mass, and $r$ is the distance between the two masses.

[6] Otávio Bueno (2019) provides a useful discussion of the status of mathematics in OSR.

Epistemic Structural Realism (ESR). Unlike explanationist realists, who think that our best scientific theories give us knowledge about unobservable entities, processes, and events (see Sect. 3.1), epistemic structural realists reserve their realist commitments to unobservable structures only. For example, Augustin-Jean Fresnel's wave theory of light made an extraordinary prediction, namely, that there will be a white spot at the center of the shadow of an opaque disc held in light diverging from a single slit. This was an extraordinary prediction at the time given the dominant understanding of light as corpuscular (that is, made of material particles) rather than wave-like.[7] Despite this predictive success of Fresnel's theory of light, it was later abandoned and replaced by James Clerk Maxwell's electromagnetic theory. Structural realists would argue that Fresnel's theory of light was not completely wrong; it did get something right, not about the *nature* of light, but rather about the *structure* of light. As John Worrall (1989, p. 117) puts it, "it seems right to say that Fresnel completely misidentified the *nature* of light, but nonetheless it is no miracle that his theory enjoyed the empirical predictive success that it did; it is no miracle because Fresnel's theory, as science later saw it, attributed to light the right *structure*" (emphasis in original).

Those who construe Structural Realism as an ontological position subscribe to the view that structures are ontologically basic. That is to say, everything that exists depends on the existence of structures and structures depend on nothing else for their existence. This view is known as Ontic Structural Realism (OSR). As James Ladyman (1998, p. 420) puts it, OSR "means taking structure to be primitive and ontologically subsistent." In that respect, OSR is a metaphysical stance, whereas ESR is an epistemic stance. As we have seen in Chap. 2 (see Sect. 2.1), the epistemic stance of scientific realism is the thesis that "Mature and predictively successful scientific theories are well-confirmed and approximately true. So entities posited by them, or, at any rate entities very similar to those posited, inhabit the world" (Psillos 2006, p. 135). ESR falls under the epistemic dimension of scientific realism, then, since it is the view that the unobservable *structures* posited by our mature and predictively successful scientific theories, or structures very similar to those posited, make up the world.

On the other hand, the metaphysical stance of scientific realism is the thesis that the "world has a definite and mind-independent structure" (Psillos 2006, p. 135). OSR falls under the metaphysical dimension of scientific realism, then, since it is the view that only *structure* is real. As French and Ladyman (2003, p. 37) point out, the motivation behind OSR is to "satisfy the 'mind independence' requirements of realism in general," that is, the metaphysical thesis, stance, or dimension of scientific realism (see Chap. 2, Sect. 2.1) but with respect to unobservable *structures* only, not unobservable entities, processes, or events.[8] Several realist arguments

---

[7] Alberto Cordero (2011) discusses this example and its implication for the selective (or *divide et impera*) realist strategy in more detail.

[8] Roman Frigg and Ioannis Votsis (2011) provide a comprehensive survey of various positions that fall under the "Structural Realism" label.

support Structural Realism, including the so-called "miracle" argument or "no miracles" argument, which is analyzed and evaluated in Chap. 4, Sect. 4.1.

## 3.6 Summary

The scientific realism/antirealism debate in contemporary philosophy of science has taken what might be called a "selectivist turn" insofar as contemporary scientific realists are more selective about the aspects of science they are willing to be realists about than previous scientific realists were. Accordingly, scientific realists of the explanationist stripe reserve their realist commitments to the theoretical posits that are responsible for--or can best explain--the predictive success of our best scientific theories (see Sect. 3.1). Entity realists argue that only those unobservable theoretical entities that can be causally manipulated and that enable efficacious interventions in nature are real (see Sect. 3.4). Structural realists, either of the ontic variety or the epistemic variety, argue that the parts of our best scientific theories that warrant a realist commitment are unobservable structures, such as those represented by relations captured in the equations of our best scientific theories, rather than unobservable entities, processes, or events (see Sect. 3.5). Like scientific realism, antirealism also comes in different varieties. For constructive empiricists, the aim of science is empirical adequacy, not (approximate) truth, and so we should believe what our best scientific theories say about observable phenomena, but we should not believe what they say about unobservables (see Sect. 3.3). For instrumentalists, scientific theories are not to be taken literally, that is, scientific theories are not the sort of things that can be true or false. Rather, scientific theories are merely instruments for attaining practical goals, such as predicting the occurrence of natural phenomena (see Sect. 3.2).

## Glossary

**Antirealism** An agnostic or skeptical attitude toward the theoretical posits (that is, unobservables) of scientific theories. Antirealism comes in different varieties, such as Constructive Empiricism (see Chap. 3, Sect. 3.3) and Instrumentalism (see Chap. 3, Sect. 3.2).

**Approximate truth** Closeness to the truth or truthlikeness. To say that a theory is approximately true is to say that it is close to the truth. According to some scientific realists, approximate truth is the aim of science. (See Chap. 2, Sect. 2.1).

**Constructive Empiricism** The view that the aim of science is to construct empirically adequate theories. A theory is empirically adequate when what the theory says about what is observable (by us) is true. (See Chap. 3, Sect. 3.3).

**Direct observation** Observation with the naked eye, without the use of scientific instruments, such as microscopes and telescopes, as opposed to instrument-aided observation. (See Chap. 3, Sect. 3.3).

**Empirical adequacy** The aim of science, according to Constructive Empiricism. To say that a theory is empirically adequate is to say that what the theory says about what is observable (by us) is true. (See Chap. 3, Sect. 3.3).

**Empirical success** A scientific theory is said to be empirically successful just in case it is both explanatorily successful (that is, it explains natural phenomena that would otherwise be mysterious to us) and predictively successful (that is, it makes predictions that are borne out by observation and experimentation). (See Chap. 3, Sect. 3.1).

**Entity Realism** The view that the theoretical entities (that is, unobservables) posited by our best scientific theories are real. (See Chap. 3, Sect. 3.4).

**The epistemic dimension (or stance) of scientific realism** The thesis that our best scientific theories, in particular, those that are empirically successful, are approximately true. (See Chap. 2, Sect. 2.1).

**Epistemic Structural Realism (ESR)** The view that the best scientific theories give us knowledge about the unobservable structure of the world. (See Chap. 3, Sect. 3.5).

**Explanationist Realism** The view that realist commitments are warranted with respect to the theoretical posits that are responsible for--or can best explain--the predictive success of our best scientific theories (also known as "Deployment Realism"). (See Chap. 3, Sect. 3.1).

**Explanatory success** A scientific theory is said to be explanatorily successful just in case it explains natural phenomena that would otherwise be mysterious to us. (See Chap. 3, Sect. 3.1).

**Instrument-aided observation** Observation by means of scientific instruments, such as microscopes and telescopes, as opposed to direct or naked-eye observation. (See Chap. 3, Sect. 3.3).

**Instrumentalism** The view that scientific theories are instruments for attaining practical goals, such as predicting the occurrence of natural phenomena. (See Chap. 3, Sect. 3.2).

**The metaphysical dimension (or stance) of scientific realism** The thesis that there are things out there in the world for scientists to discover and that those things out there in the world are independent of the human minds that study them. (See Chap. 2, Sect. 2.1).

**Modus ponens** A form of argument with a conditional premise, a premise that asserts the antecedent of the conditional premise, and a conclusion that asserts the consequent of the conditional premise. That is, "if $A$, then $B$, $A$; therefore, $B$," where $A$ and $B$ stand for statements. *Modus ponens* is a valid form of inference, and so an argument in natural language that takes this logical form is valid. On the other hand, the following logical form is invalid: "if $A$, then $B$, $B$; therefore, $A$." It is known as the fallacy of affirming the consequent. (See Chap. 4, Sect. 4.1).

**Modus tollens** A form of argument with a conditional premise, a premise that denies the consequent of the conditional premise, and a conclusion that denies the antecedent of the conditional premise. That is, "if *A*, then *B*, not *B*; therefore, not *A*," where *A* and *B* stand for statements. *Modus tollens* is a valid form of inference, and so an argument in natural language that takes this logical form is valid. On the other hand, the following logical form is invalid: "if *A*, then *B*, not *A*; therefore, not *B*." It is known as the fallacy of denying the antecedent. (See Chap. 5, Sect. 5.1).

**Ontic Structural Realism (OSR)** The view that everything that exists depends on the existence of structures and structures depend on nothing else for their existence. (See Chap. 3, Sect. 3.5).

**Predictive success** A scientific theory is said to be predictively successful just in case it makes predictions that are borne out by observation and experimentation. (See Chap. 3, Sect. 3.1).

**Scientific realism** An epistemically positive attitude toward those aspects of scientific theories that are worthy of belief. Scientific realism comes in different varieties, such as Explanationist Realism (see Chap. 3, Sect. 3.1), Entity Realism (see Chap. 3, Sect. 3.4), Structural Realism (see Chap. 3, Sect. 3.5), and Relative Realism (see Chap. 6, Sect. 6.1).

**The semantic dimension (or stance) of scientific realism** The thesis that scientific theories are to be taken literally, which means that they can be either true or false. (See Chap. 2, Sect. 2.1).

## References and Further Readings

Abbott, B. P., et al. (2016). Observation of gravitational waves from a binary black hole merger. *Physical Review Letters, 116*(061102), 1–16.

Bueno, O. (2019). Structural realism, mathematics, and ontology. *Studies in History and Philosophy of Science Part A, 74*(2019), 4–9.

Castelvecchi, D. (2019). Black hole imaged for first time. *Nature, 568,* 284–285.

Cordero, A. (2011). Scientific realism and the divide et impera strategy: The ether saga revisited. *Philosophy of Science, 78*(5), 1120–1130.

Duhem, P. (1954/1982). *The aim and structure of physical theory.* Princeton, NJ: Princeton University Press. Translated from the French by Philip P. Wiener.

French, S. (2014). *The structure of the world: Metaphysics and representation.* Oxford: Oxford University Press.

French, S., & Ladyman, J. (2003). Remodelling structural realism: quantum physics and the metaphysics of structure. *Synthese, 136*(1), 31–56.

Frigg, R., & Votsis, I. (2011). Everything you always wanted to know about structural realism but were afraid to ask. *European Journal for Philosophy of Science, 1*(2), 227–276.

Galileo, G. (1623/1957). The Assayer. In S. Drake (Ed.), *Discoveries and opinions of Galileo* (pp. 231–280). New York: Anchor Books.

Hacking, I. (1983). *Representing and intervening: Introductory topics in the philosophy of natural science.* New York: Cambridge University Press.

Hacking, I. (1989). Extragalactic reality: The case of gravitational lensing. *Philosophy of Science, 56*(4), 555–581.

Ivanova, M. (2013). Did Perrin's experiments convert Poincaré to scientific realism? *HOPOS: The Journal of the International Society for the History of Philosophy of Science, 3*(1), 1–19.

Kitcher, P. (1989). Explanatory unification and the causal structure of the word. In P. Kitcher & W. Salmon (Eds.), *Scientific explanation* (pp. 410–505). Minneapolis: University of Minnesota Press.

Kitcher, P. (1993). *The advancement of science: Science without legend, objectivity without illusions*. New York: Oxford University Press.

Kitcher, P. (2002). Scientific knowledge. In P. K. Moser (Ed.), *The Oxford handbook of epistemology* (pp. 385–407). New York: Oxford University Press.

Ladyman, J. (1998). What is structural realism? *Studies in History and Philosophy of Science Part A, 29*(3), 409–424.

Land, M. F., & Nilsson, D.-E. (2012). *Animal eyes* (Oxford animal biology series) (2nd ed.). New York: Oxford University Press.

Lyons, T. D. (2016). Structural realism versus deployment realism: A comparative evaluation. *Studies in History and Philosophy of Science Part A, 59*, 95–105.

Mizrahi, M. (2014). Constructive empiricism: normative or descriptive? *International Journal of Philosophical Studies, 22*(4), 604–616.

Nanay, B. (2019). Entity realism and singularist semirealism. *Synthese, 196*(2), 499–517.

Overbye, D. (2019). Darkness visible, finally: Astronomers capture first ever image of a Black Hole. *The New York Times*, April 10.

Psillos, S. (1999). *Scientific realism: How science tracks truth*. London: Routledge.

Psillos, S. (2006). Thinking about the ultimate argument for realism. In C. Cheyne & J. Worrall (Eds.), *Rationality and reality: Conversations with Alan Musgrave* (pp. 133–156). Dordrecht: Springer.

Putnam, H. (1994). In J. Conant (Ed.), *Words and life*. Cambridge, MA: Harvard University Press.

Rowbottom, D. P. (2019). *The instrument of science: Scientific anti-realism revitalised*. London: Routledge.

Russell, B. (1912). *The problems of philosophy*. New York: Henry Holt and Company.

Sankey, H. (2008). *Scientific realism and the rationality of science*. Hampshire: Ashgate.

Shapere, D. (1993). Astronomy and antirealism. *Philosophy of Science, 60*(1), 134–150.

van Fraassen, B. C. (1980). *The scientific image*. New York: Oxford University Press.

van Fraassen, B. C. (1998). The agnostic subtly probabilified. *Analysis, 58*(3), 212–220.

van Fraassen, B. C. (2001). Constructive empiricism now. *Philosophical Studies, 106*(1–2), 151–170.

Worrall, J. (1989). Structural realism: the best of both worlds? *Dialectica, 43*(1–2), 99–124.

# Chapter 4
# Key Arguments for Scientific Realism

**Abstract** In this chapter, I present in canonical (or standard) form and then evaluate key arguments for scientific realism (or against antirealism about science). The first argument is the most influential Positive Argument for scientific realism, most commonly known as the "no miracles" argument, which purports to show that scientific realism is the best explanation for the empirical success of science. The second argument, which can be found in Grover Maxwell's seminal (1962) paper, purports to show that there is no principled distinction between observables and unobservables, and thus, if our belief in the existence of the former is justified, which it is, then our belief in the existence of the latter is justified as well. The third argument can also be found in Maxwell's seminal (1962) paper and it purports to show that belief in the existence of the theoretical entities of our best scientific theories, even if they are unobservable, is justified because any theoretical entity can become observable *in principle*. The fourth argument builds on Ian Hacking's (Representing and intervening: introductory topics in the philosophy of natural science. Cambridge University Press, New York, 1983, p. 23) famous slogan for Entity Realism, "if you can spray them, they are real," and proceeds from corroborating the existence of theoretical entities through results obtained by distinct experimental means. The fifth argument is based on Ludwig Fahrbach's (Synthese 180(2):139–155, 2011) observation that science is growing at an exponential rate. When weighted exponentially, the best scientific theories since the 1950s appear to have been stable, which in turn inspires an optimistic induction to the conclusion that our best scientific theories will remain stable.

**Keywords** Corroboration · Direct observation · Exponential growth of science · Historical graveyard of science · Inference to the Best Explanation (IBE) · Instrument-aided observation · Empirical success · Observable/unobservable distinction · Observation/detection distinction · Predictive success · Public hallucination · Selectionist explanation of success · Theoretical virtues · Theory-laden · Underdetermination of theories by evidence

In this chapter, I present in canonical (or standard) form and then evaluate key arguments for scientific realism (or against antirealism about science). The first argument is the most influential Positive Argument for scientific realism, most commonly

© Springer Nature Switzerland AG 2020
M. Mizrahi, *The Relativity of Theory*, Synthese Library 431,
https://doi.org/10.1007/978-3-030-58047-6_4

known as the "no miracles" argument, which purports to show that scientific realism is the best explanation for the empirical success of science. The second argument, which can be found in Grover Maxwell's seminal (1962) paper, purports to show that there is no principled distinction between observables and unobservables, and thus, if our belief in the existence of the former is justified, which it is, then our belief in the existence of the latter is justified as well. The third argument can also be found in Maxwell's seminal (1962) paper and it purports to show that belief in the existence of the theoretical entities of our best scientific theories, even if they are unobservable, is justified because any theoretical entity can become observable *in principle*. The fourth argument builds on Ian Hacking's (1983, p. 23) famous slogan for Entity Realism, "if you can spray them, they are real," and proceeds from corroborating the existence of theoretical entities through results obtained by distinct experimental means. The fifth argument is based on Ludwig Fahrbach's (2011) observation that science is growing at an exponential rate. When weighted exponentially, the best scientific theories since the 1950s appear to have been stable, which in turn inspires an optimistic induction to the conclusion that our best scientific theories will remain stable.

## 4.1  The Positive Argument for Scientific Realism

One of the key arguments for scientific realism is known as the "miracle" argument or the "no miracles" argument. This argument purports to show that scientific realism is the best explanation for the fact that science is successful. When philosophers of science talk about "the success of science," they usually mean empirical success, which includes explanatory and predictive success. That is to say, the best scientific theories are those that explain natural phenomena that would otherwise be mysterious to us and that make predictions that are borne out by observation and experimentation. As far as the so-called "no miracles" argument goes, the latter kind of success, namely, *predictive success*, is particularly important. For, as the argument goes, it would be a miracle if our best scientific theories were to make predictions that are borne out by observational and experimental testing and yet those theories were not even approximately true. Hilary Putnam succinctly made this argument for scientific realism in his (1975, p. 73).

Putnam, Hilary. 1975. *Mathematics, matter and method*. New York: Cambridge University Press.

*[T]he positive argument for scientific realism is that it is the only philosophy that doesn't make the success of science a miracle.*

As I understand it, the argument that Putnam is making in this passage can be stated in canonical (or standard) form as follows:

(P1) Our best scientific theories successfully explain natural phenomena and make predictions that are confirmed by observations and experimental testing.

(P2) The best explanation for the phenomenon described in (P1) is scientific realism, that is, the view that our best scientific theories are approximately true.

(P3) No other explanation explains the phenomenon described in (P1) as well as scientific realism does.

Therefore,

(C) Our best (that is, empirically successful) scientific theories are approximately true.[1]

This Positive Argument for scientific realism is an instance of what philosophers of science call "Inference to the Best Explanation" (IBE).[2] As James Ladyman (2002, p. 209) points out, IBE "is sometimes also known as 'abduction' – following the terminology of Charles Peirce." However, some philosophers have argued that IBE and Peirce's abduction are different forms of inference (Douven 2017).[3] Be that as it may, for Peirce, abduction is a non-deductive form of inference. Likewise, IBE is typically construed as an ampliative, or non-deductive, form of argumentation that proceeds from a phenomenon that requires an explanation to the conclusion that the best explanation for that phenomenon is probably true.[4] As Ladyman (2007, p. 341) describes it, "Inference to the best explanation (IBE) is a (putative) rule of inference according to which, where we have a range of competing hypotheses all of which are empirically adequate to the phenomena in some domain, we should infer the truth of the hypothesis which gives us the best explanation of those phenomena." The general form of IBE can be stated as follows:

1. Phenomenon $P$.
2. The best explanation for $P$ is $E$.
3. No other explanation explains $P$ as well as $E$ does.
4. Therefore, (probably) $E$.[5]

In the case of the Positive Argument for scientific realism, the phenomenon that requires an explanation is the empirical success of our best scientific theories, which

---

[1] Notice that Putnam's claim is that scientific realism is the *only* explanation for the success of science. Strictly speaking, then, Putnam's Positive Argument seems to be an instance of what Alexander Bird (2007, p. 425) calls "Inference to the Only Explanation" (IOE), which is an inference to "the truth of some hypothesis since it is the only possible hypothesis left unrefuted by the evidence. It is the form of inference advocated by Sherlock Holmes in his famous dictum 'Eliminate the impossible, and whatever remains, however improbable, must be the truth'." In that case, however, the premise according to which scientific realism is the *only* explanation for the success of science would be false, given that there are alternative, antirealist explanations for the empirical success of science, as we will see. For this reason, it seems better (that is, more charitable) to interpret Putnam's Positive Argument as an IBE rather than an IOE.

[2] The phrase was coined by Gilbert Harman (1965).

[3] Daniel Campos (2011, p. 419) argues against the "tendency in the philosophy of science literature to link abduction to the inference to the best explanation (IBE), and in particular, to claim that Peircean abduction is a conceptual predecessor to IBE."

[4] For example, Alan Baker (2010, pp. 37-38) defines IBE as "A method of reasoning, also known as *abduction*, in which the truth of an hypothesis is inferred on the grounds that it provides the best explanation of the relevant evidence. In general, inference to the best explanation (IBE) is an ampliative (i.e., non-deductive) method" (emphasis in original).

[5] For more on the structure of IBE, see Psillos (2007).

includes explanatory success and, most importantly, predictive success (see Chap. 3, Sect. 3.1). This empirical success is supposed to be explained by the hypothesis that successful scientific theories are approximately true, that is, by scientific realism. If scientific realism does indeed provide the best explanation for the explanatory and predictive success of our best scientific theories, then that would be a good reason to believe that scientific realism is probably correct. In other words, the premises of the Positive Argument for scientific realism, if true, would provide strong probable (but not conclusive) support for its conclusion. Given that the Positive Argument for scientific realism is not meant to be a deductive argument, the next question is not whether it is valid or invalid, but rather whether it is strong or weak.

Antirealists have objected both to the premises of the Positive Argument for scientific realism and to the reasoning itself. In particular, they argue that IBE is not a truth conducive form of argumentation. For example, according to Bas van Fraassen (1980), the Positive Argument for scientific realism is fallacious. From the fact that an explanation is the best one we could come up with, it does not follow that the explanation is probably true. For we might be working with a bad lot of explanations and the likely true explanation simply did not occur to us. As van Fraassen (1980, p. 143) puts it:

> [IBE] is a rule that selects the best among the historically given hypotheses. We can watch no contest of the theories we have so painfully struggled to formulate, with those no one has proposed. So *our selection may well be the best of a bad lot* (emphasis added).

According to van Fraassen, then, whenever scientists infer that one explanation is better than alternative explanations for the same phenomenon on explanatory grounds, they may simply be selecting the best explanation among several bad ones. If van Fraassen is right about this, then any inference to the best explanation would be a weak argument. For, even if no other explanation explains $P$ (for example, the empirical success of our best scientific theories) as well as $E$ (for example, scientific realism) does, $E$ would still not be more probable or likely to be true if $E$ is simply the best of a "bad lot" of explanations.[6]

Furthermore, van Fraassen also argues that explanations are judged as the *best* based on criteria that are not truth conducive. That is to say, if we have several competing explanations for the same phenomenon, how do we rank them from best to worst? We need selection criteria. Philosophers of science have offered criteria for selecting the best explanation among several competing explanations. These selection criteria are properties of theories that are considered desirable, that is, they are properties that make theories that have them good theories. For this reason, these properties of theories are also called "theoretical virtues." According to James Ladyman (2007, p. 340), for instance, properties of theories that are deemed "virtuous" include "simplicity, non-ad hocness, novel predictive power, elegance, and explanatory power."[7] Such theoretical virtues give scientists reasons to choose a

---

[6] We will revisit van Fraassen's criticism against IBE as inference from a "bad lot" of explanations in Chap. 5, Sect. 5.2, and Chap. 6, Sect. 6.4.

[7] When philosophers of science talk about "novel predictions," they typically mean a prediction that was not known to be true (or was expected to be true or false) at the time the theory was constructed.

theory (a hypothesis or an explanation) among several competing theories. For our purposes, the following partial (and by no means exhaustive) list of criteria of selection (or theoretical virtues) for competing explanations is sufficient (Mizrahi 2012, p. 134):

> *Unification*: As a general rule of thumb, choose the explanation (or hypothesis or theory) that explains the most and leaves the least unexplained things.
> *Testability*: As a general rule of thumb, choose the explanation (or hypothesis or theory) that yields independently testable predictions.
> *Coherence*: As a general rule of thumb, choose the explanation (or hypothesis or theory) that is consistent with background knowledge.
> *Simplicity*: As a general rule of thumb, choose the least complicated explanation (or hypothesis or theory), i.e., the one that posits the least causal sequences and entities, and that goes beyond the evidence the least.[8]

For example, a certain explanation might be simpler than other explanations for the same phenomenon, and thus it might be judged to be the best one. But simplicity and the like are pragmatic virtues, rather than epistemic virtues, of explanations. That is to say, the fact that an explanation is simpler than other explanations for the same phenomenon does not necessarily mean that it is the correct one. This is because there is no reason to think that the world is simple. In fact, van Fraassen argues, it is absurd to believe that the world is more likely to be simple than complicated. As van Fraassen (1980, p. 90) puts it, "it is surely absurd to think that the world is more likely to be simple than complicated (unless one had certain metaphysics or theological views not usually accepted as legitimate factors in scientific inference)." If van Fraassen is right about this, then the fact that one explanation is simpler than alternative explanation for the same phenomenon does not make that explanation more likely to be true than the other explanations. The same point is supposed to apply to the other theoretical virtues because they are all pragmatic, not epistemic, virtues. That is to say, like simplicity, unification, coherence, and the rest of the selection criteria are merely pragmatic virtues; a theory that has these properties is not more likely to be true than a theory that lacks them. So, even if scientific realism is the best explanation for the empirical success of our best scientific theories, as scientific realists argue, antirealists like van Fraassen would insist that the inference to the likely truth of scientific realism is unlicensed. For these reasons, the Positive Argument (also known as the "no miracles" argument) cannot be said to be a strong argument for scientific realism.[9]

---

[8] Michael Keas (2018) provides a comprehensive list of theoretical virtues, which includes accuracy, causal adequacy, explanatory depth, consistency, coherence, beauty, simplicity, unification, durability, fruitfulness, and applicability. Samuel Schindler's (2018) book on theoretical virtues is a useful, comprehensive study of theoretical virtues in science. In Mizrahi (2020), I report the results of a corpus-based, empirical study of theoretical virtues in scientific practice.

[9] As we will see in Chap. 5, Sect. 5.2, even though antirealists (specifically, constructive empiricists) reject IBE as an illegitimate form of inference in science, they sometimes fall back on IBE when giving a "positive argument" for their own antirealist position (specifically, Constructive Empiricism). See Mizrahi (2018).

Antirealists tend to find IBE problematic in general, but they have also argued that the Positive Argument for scientific realism is especially problematic in particular. This is because the so-called "no miracles" argument is circular, or so some antirealists argue. An argument is circular when it assumes as a premise the very conclusion it is intended to be an argument for. This is problematic, as far as arguments go, because such an argument would fail to persuade anyone who does not already accept the conclusion of the argument. This is why circular reasoning is also called "begging the question," that is, assuming an answer to a question that needs to be argued for rather than taken for granted as a premise in an argument.[10] When it comes to the Positive Argument for scientific realism, some antirealists have argued that scientific realists are begging the question against antirealism when they argue that approximate truth is the best explanation of the empirical success of our best scientific theories. For the very question at issue is whether the empirical success of a theory is a good reason to believe that the theory is approximately true, and this is a question that antirealists tend to answer in the negative because, as we have seen (see Chap. 2, Sect. 2.1), antirealists generally reject the possibility of theoretical knowledge in science. Here is how Larry Laudan (1981, p. 45) puts the charge of circularity against the Positive Argument for scientific realism:

> Now enters the new breed of realist [for example, Hilary Putnam] who wants to argue that epistemic realism [that is, the epistemic thesis of scientific realism, see Chapter 2, Sect. 2.1] can reasonably be presumed to be true by virtue of the fact that it has true consequences. But this is a monumental case of begging the question. The non-realist refuses to admit that a *scientific* theory can be warrantedly judged to be true simply because it has some true consequences [that is, it is empirically successful]. Such non-realists are not likely to be impressed by the claim that a *philosophical* theory like realism can be warranted as true because it arguably has some true consequences. If non-realists are chary about first-order abductions [that is, IBEs] to avowedly true conclusions, they are not likely to be impressed by second-order abductions [that is, IBEs], particularly when, as I have tried to show above, the premises and conclusions are so indeterminate (emphasis in original).[11]

In other words, it is question-begging, according to Laudan, to apply the rule of IBE to philosophical theories about scientific theories, such as scientific realism, when antirealists do not accept that the rule even applies to scientific theories in the first place. As James Ladyman et al. (2007, p. 75) put it, "since it is the use of IBE in specific instances where this involves unobservables that is in question, it is viciously circular to use IBE at the global level to infer the truth of scientific realism because the latter is a hypothesis involving unobservables." That is to say, if IBE does not yield theoretical knowledge in science, as antirealists would argue, then it cannot yield theoretical knowledge in philosophy of science, either.

---

[10] As Stathis Psillos (2011, p. 25) puts it, "Vicious circularity is an epistemic charge – a viciously circular argument has no epistemic force. It cannot offer reasons to believe the conclusion. It cannot be persuasive" (emphasis in original).

[11] Similarly, Arthur Fine argues that scientific realists are "not free to assume the validity of a principle [namely, IBE] whose validity is itself under debate" (1986, p. 161) and that the so-called "no miracles" argument for scientific realism is viciously circular (or question-begging) because it is "the very type of argument whose cogency is the question under discussion" (1991, p. 82).

In response to this charge of circularity, scientific realists have argued that the Positive Argument for scientific realism may be circular but not *viciously* circular. For example, Stathis Psillos (1999, p. 82), following R. B. Braithwaite (1953), distinguishes between premise-circular arguments, in which "one claims to offer an argument for the truth of α, but explicitly *presupposes* α in one's premises," and rule-circular arguments, in which "the argument itself is an instance of, or involves essentially an application of, the rule of inference vindicated by the conclusion" (1999, p. 82). Psillos argues that premise-circular arguments are viciously circular, whereas rule-circular arguments are not viciously circular. Since the so-called "no miracles" argument for scientific realism is rule-circular, but not premise-circular, it follows that it is not viciously circular.

However, recall that an explanation is considered the best among several competing explanations for the same phenomena just in case it has at least some of the properties that make explanations good explanations. As we have seen, some of the good-making properties of explanations include such properties (or theoretical virtues) as unification, testability, coherence, and simplicity. In other words, the best explanation for some phenomenon is the explanation that unifies that which needs to be explained (that is, explains the most and leaves the least unexplained), makes testable predictions (that is, has consequences that can be observed to be the case independently of whether the explanation itself is true or false), is coherent (that is, is consistent with our background knowledge about the phenomenon in question), and is simple (that is, does not go far beyond the evidence in positing complex entities, processes, or events). To put it another way, the term "best," as it is used in IBE, means "the most unifying, testable, coherent, and simple" explanation. With that in mind, let us substitute the term "best" wherever it occurs in the Positive Argument for scientific realism with this precise definition. With this substitution, the Positive Argument for scientific realism now runs as follows:

(P1) Our most unifying, testable, coherent, and simple scientific theories successfully explain natural phenomena and make predictions that are confirmed by observations and experimental testing.

(P2) The most unifying, testable, coherent, and simple explanation for the phenomenon described in (P1) is scientific realism, that is, the view that our most unifying, testable, coherent, and simple scientific theories are approximately true.

(P3) No other explanation explains the phenomenon described in (P1) as well as scientific realism does.

Therefore,

(C) Our most unifying, testable, coherent, and simple scientific theories are approximately true.

Now it appears that there is something borderline circular going on in the Positive Argument for scientific realism. With the definition of "best" in terms of theoretical virtues, particularly the theoretical virtue of testability, (P1) amounts to saying that our most predictively successful scientific theories make successful predictions, which seems rather tautological. Obviously, the most predictively successful theories are predictively successful. This is not an instance of circular reasoning per se. Yet, it might still be considered an instance of a circular definition. A circular definition is a definition that uses the very term that it is supposed to define in the definition itself. In (P1), we are simply told that testable theories are testable. If this is correct, then the so-called "no miracles" argument turns on a circular definition of "best" because it amounts to arguing that scientific realism is the most predictively successful explanation for why our most predictively successful theories are the most predictively successful.

Likewise, with the definition of "best" in terms of theoretical virtues, particularly the theoretical virtue of testability, (P2) amounts to saying that the most predictively successful explanation for predictive success is scientific realism. But this is a premise that antirealists would not accept. Nor is (P3) a premise that antirealists would accept (or even could be reasonably expected to accept). For they have offered alternative explanations for the predictive success of our best scientific theories. For example, the fact that our best scientific theories make accurate predictions can be explained by an appeal to a selection process akin to a Darwinian process of natural selection. As Brad Wray (2007, p. 84) puts it:

> any theory that does not enable us to make accurate predictions is not apt to be around very long. No scientist will waste her career working with such a theory. As a result, *any theory that is still around is apt to be successful.* Consequently, when philosophers of science look at the world of science they should not be surprised to find only successful theories. The others have been eliminated or are on their way to being eliminated (emphasis added).

That is to say, scientific theories that fail to make accurate predictions are weeded out and replaced by those that succeed in doing so. This selection process explains why our best scientific theories are predictively successful for, if they were not predictively successful, this selection process would have eliminated them. This selectionist explanation for the empirical success of science, antirealists argue, explains the predictive success of our best scientific theories just as well as scientific realism does.[12]

In fact, antirealists would argue that the selectionist explanation is better than scientific realism, for it explains something that scientific realism cannot explain. According to Wray, the selectionist explanation explains not only the empirical success of science but also the empirical failures of science. According to the selectionist explanation, there is a selection process of theories in science that is analogous to the selection process of species in nature. According to the selectionist explanation, scientific theories that fail to make accurate predictions are weeded out and replaced by those that succeed in doing so. This selectionist explanation would explain not only the empirical success of theories that survive, for if they were not empirically successful, this selection process would have eliminated them, but also

---

[12] Kyle Stanford (2000) offers another antirealist explanation of the success of science in terms of "predictive similarity." According to Stanford, the predictive success of an abandoned theory can be explained by pointing out how closely its predictions approximate those of the accepted theory. For example, we "explain the success of the (revised) Ptolemaic system of epicycles by pointing out how closely its predictions approximate those of the true Copernican hypothesis. Let us call this relationship the *predictive similarity* of the Ptolemaic system to the Copernican" (Stanford 2000, p. 273).

the extinction of empirically unsuccessful theories, that is, those that were eliminated by the selection process. The latter is a phenomenon that scientific realism
cannot explain. That is to say, just as there are scientific theories that are empirically
successful, there are also scientific theories that are unsuccessful. Scientific realism
can explain the empirical success of science, but it cannot explain the failures. Of
course, that a scientific theory is approximately true would not explain why it is
empirically unsuccessful. For, if it were approximately true, we would expect that
theory to be empirically successful, not unsuccessful. As Wray (2018, p. 150) puts
it, the selectionist explanation "provides us with the resources to explain the *failures*
of science. The realist seems to have nothing to say here" (emphasis in original).[13]

Accordingly, if there are alternative explanations to the empirical success of our
best scientific theories, such as the selectionist explanation, which explain the
empirical success of our best scientific theories just as well as scientific realism
does, if not better, then (P3) of the Positive Argument for scientific realism is false.
If (P3) is false, then the Positive Argument for scientific realism (also known as the
"no miracles" argument), even if it were a strong argument, antirealist objections to
IBE notwithstanding, cannot be said to be a cogent argument for scientific realism.

Although most contemporary scientific realists understand the Positive Argument
for scientific realism as an Inference to the Best Explanation (IBE), and IBE as "an
ampliative (i.e., non-deductive)" form of inference (Baker 2010, p. 38), it is important to note that there were a few scientific realists for whom the so-called "no
miracles" argument for scientific realism was best construed as a deductive argument. For instance, Alan Musgrave (1988) subscribed to deductivism, which is the
view that ampliative (or non-deductive) arguments are really incomplete deductive
arguments. Such incomplete arguments, or enthymemes, can (and should) be made
fully deductive when their missing premises are made explicit. For Musgrave (1988,
p. 239), then, the missing premise in any argument that appears to take the form of
IBE is the following: "It is reasonable to accept a *satisfactory* explanation of any
fact, which is also the best available explanation of that fact, as true" (emphasis in
original). With this implicit premise, now made explicit, any argument that appears
to take the form of IBE can (and should) be turned into a deductive argument as
follows (Musgrave 1988, p. 239):

1. It is reasonable to accept a *satisfactory* explanation of any fact, which is also the
   best available explanation of that fact, as true.
2. *F* is a fact.
3. Hypothesis *H* explains *F*.
4. Hypothesis *H satisfactorily* explains *F*.
5. No available competing hypothesis explains *F* as well as *H* does.
6. Therefore, it is reasonable to accept *H* as true (emphasis in original).

With the addition of premise 1, Musgrave argues, an argument that appears to be an
ampliative (or non-deductive) argument now becomes a deductive argument. This

---

[13] We will come back to the selectionist explanation in Chap. 6 (see Sect. 6.5).

can be seen more clearly, perhaps, in the way Stathis Psillos (2006, p. 140) reconstructs Musgrave's deductive version of IBE:

1. If hypothesis *H* is the best explanation of the fact to be explained, then it is reasonable to accept *H* as true.
2. *H* is the best explanation of the evidence.
3. Therefore, it is reasonable to accept *H* as true.

This is a deductive argument because the premises purport to provide logically conclusive support, rather than probable support, for the conclusion. Now, when the hypothesis in question is scientific realism and the fact to be explained is the empirical success of our best scientific theories, Musgrave's deductive version of the Positive Argument for scientific realism can be stated as follows:

> (P1) If scientific realism is the best explanation of the fact that our best scientific theories are empirically successful, then it is reasonable to accept scientific realism as true.
> (P2) Scientific realism is the best explanation of the fact that our best scientific theories are empirically successful.
>
> Therefore,
>
> (C) It is reasonable to accept scientific realism as true.

As stated, this version of the Positive Argument for scientific realism is a deductive argument (in particular, a *modus ponens*: If *A*, then *B*, *A*; therefore, *B*). Since the premises of this argument, namely, (P1) and (P2), successfully provide logically conclusive support for the conclusion, namely, (C), this argument can be said to be valid. The next question, then, is whether the premises are in fact true. Is Musgrave's deductive version of the Positive Argument a sound argument for Relative Realism?

As we have already seen, antirealists would not accept (P2). For they have offered competing hypotheses to explain the predictive success of our best scientific theories, such as the selectionist explanation. In fact, as we discussed above, antirealists would argue that the selectionist explanation is better than scientific realism, for it explains not only the successes but also the failures of science. Accordingly, if there are alternative explanations to the empirical success of our best scientific theories, such as the selectionist explanation, which explain the empirical success of our best scientific theories just as well as scientific realism does, if not better, then (P2) of Musgrave's deductive version of the Positive Argument for scientific realism is false. If (P2) is false, then Musgrave's deductive version of the Positive Argument for scientific realism, albeit valid, cannot be said to be a sound argument for scientific realism.

Scientific realists could object that the selectionist explanation does not really explain the empirical success of any given theory. Instead, it merely explains why there are few (or no) unsuccessful theories in current science. Indeed, Musgrave himself has made this objection against van Fraassen's selectionist explanation. As Musgrave (1988, p. 242) writes:

> It is one thing to explain why only successful theories survive, and quite another thing to explain why some particular theory is successful. Van Fraassen's Darwinian explanation of the former can be accepted by realist and anti-realist alike. But it yields no explanation at

all of the latter. You do not explain why (say) electron-theory is (scientifically) successful by saying that if it had not been it would have been eliminated. Just as you do not explain why (say) the mouse is (biologically) successful by saying that if it has not been it would have been eliminated. Biologists explain why the mouse is successful by telling a long story about its well-adaptedness. Realists want to explain why electron-theory is successful by telling a shorter story about its 'well-adaptedness', that is, its truth.

In other words, just as a biological species has certain traits that make it well-adapted to its environment, and thus able to survive and reproduce, a successful theory must have certain traits that make it empirically successful. The trait that makes successful theories empirically successful, according to Musgrave, is (approximate) truth. According to Musgrave, antirealists need to point to some trait of successful theories, other than (approximate) truth, that makes them empirically successful.

Constructive empiricists would respond to Musgrave's objection by pointing to empirical adequacy. That is to say, the trait that makes successful theories empirically successful, according to constructive empiricists, is empirical adequacy. Since all empirically adequate theories are empirically successful, any given empirically adequate theory is a successful theory. Musgrave would not be impressed by an explanation that appeals to empirical adequacy. For him, "This is like explaining why some crows are black by saying that they all are" (Musgrave 1988, p. 242). However, according to some philosophical accounts of scientific explanation, to explain is to subsume under a generalization. For instance, according to the Deductive-Nomological model of scientific explanation (Hempel 1966, pp. 50–51), an explanation takes the form of a sound deductive argument in which the fact (or phenomenon) to be explained follows as a conclusion from the premises in the argument, one of which must be a law-like generalization, or a so-called "law of nature," without which the argument would not be valid. Likewise, according to Philip Kitcher (1989, p. 423), "Science advances our understanding of nature by showing us how to *derive* descriptions of many phenomena, using the same *pattern of derivation* again and again, and in demonstrating this, it teaches us how to reduce the number of facts we have to accept as ultimate" (emphasis added).

To illustrate with a rather simplified example, suppose that "All crows are black" is a general "law of nature." In that case, if scientific explanations are deductive arguments, then the following deductive argument explains why some particular crow is black:

1. All crows are black.
2. *c* is a crow.
3. Therefore, *c* is black.

In other words, the general law according to which all crows are black explains why some particular crow is black. For to explain is to derive descriptions of various phenomena by using deductive arguments. If Kitcher (1989, p. 448) is right that "all explanation is deductive," then, contrary to Musgrave's objection, constructive empiricists can explain the empirical success of a particular theory by appealing to the empirical adequacy of all successful theories. Such an explanation for why some particular theory is successful would run as follows:

1. All empirically adequate theories are empirically successful.
2. *T* is an empirically adequate theory.
3. Therefore, *T* is empirically successful.

Just as, for Musgrave, the explanation for the empirical success of some particular theory would run as follows:

1. All (approximately) true theories are empirically successful.
2. *T* is a (approximately) true theory.
3. Therefore, *T* is empirically successful.

This is not to say that Kitcher (1989, p. 448) is in fact right that "all explanation is deductive," or that the Deductive-Nomological model is the correct model of scientific explanation. Rather, it is merely to say that it is not obviously wrong for constructive empiricists to appeal to empirical adequacy as an explanation for why some particular scientific theory is empirically successful. It "is like explaining why some crows are black by saying that they all are," just as Musgrave (1988, p. 242) points out. But, if Kitcher is right about scientific explanation, then there is nothing obviously wrong with that.[14]

## 4.2 The Slippery Slope Argument for Instrumental Observation

One of the central issues in the scientific realism/antirealism debate in contemporary philosophy of science is whether belief in the theoretical posits of our best scientific theories, such as electrons, genes, and viruses, is justified. After all, these theoretical entities are unobservable, that is, they cannot be directly observed with the naked eye, but rather can only be detected by means of sophisticated instruments, such as cloud chambers, radio telescopes, particle accelerators, electron microscopes, and the like. But then how do we know that these scientific instruments reveal the true nature of reality? Perhaps what we observe through microscopes and telescopes are technical artifacts, that is, things that are not really there but rather are manufactured or produced by the act of observing through the instrument itself. For example, some images of stars show bright spikes emanating from stars in a cross-like pattern. These are called diffraction spikes and they are not really part of the stars themselves. Rather, they are imaging artifacts that are produced by the curved mirrors of refracting telescopes. Perhaps everything we "see" through a telescope is an imaging artifact. Can we be confident that what we see

---

[14] We will revisit the question of how to explain the empirical success of our best scientific theories in Chap. 6, Sect. 6.6.

through a telescope is really there in much the same way that we can be confident
that what we see with our own eyes is really there?

One way to argue in favor of scientific realism, then, is to show that belief in the
theoretical posits of our best scientific theories that can be observed by means of
instruments is justified just as belief in what can be observed without instruments is
justified. Such an argument for scientific realism can be found in a seminal paper by
Grover Maxwell (1962, p. 7).

> Maxwell, Grover. 1962. The ontological status of theoretical entities. In *Scientific
> Explanation, Space, and Time: Minnesota Studies in the Philosophy of Science*, eds.,
> H. Feigl and G. Maxwell, 181–192. Minneapolis: University of Minnesota Press.

> *there is, in principle, a continuous series beginning with looking through a vacuum and
> containing these as members: looking through a window-pane, looking through glasses,
> looking through binoculars, looking through a low-power microscope, looking through a
> high-power microscope, etc., in the order given. The important consequence is that, so far,
> we are left without criteria which would enable us to draw a non-arbitrary line between
> "observation" and "theory".*

As I understand it, the argument that Maxwell is making in this passage can be
stated in canonical (or standard) form as follows:

(P1) There is a continuum between observing without instruments and observing with
   instruments (that is, from observing through windows and spectacles all the way to
   observing through microscopes and telescopes).
(P2) In terms of that continuum, observing without instruments and observing with instruments differ only in degree, not in kind.

Therefore,

(C1) No principled distinction can be drawn between observing without instruments and
   observing with instruments. [from (P1) & (P2)]
(P3) If no principled distinction can be drawn between observing without instruments and
   observing with instruments, then if belief in the existence of entities that are observable without instruments is justified, then belief in the existence of entities that are
   observable with instruments is justified as well.

Therefore,

(C2) If belief in the existence of entities that are observable without instruments is justified,
   then belief in the existence of entities that are observable with instruments is justified
   as well. [from (C1) & (P3)]
(P4) Belief in the existence of entities that are observable without instruments is justified.

Therefore,

(C3) Belief in the existence of entities that are observable with instruments is justified.
   [from (C2) & (P4)]

This line of reasoning is valid. That is to say, at each step, the premises provide logically conclusive support for the conclusion that follows from those premises. The
next question, then, is whether the premises are actually true. Is Maxwell's Slippery
Slope Argument for Instrumental Observation sound?

As we have seen in Chap. 2, unlike scientific realists, antirealists tend to think
that belief in the theoretical posits of our best scientific theories is unjustified precisely because those theoretical entities are unobservable. Constructive empiricists,

for example, would insist that the theoretical entities, processes, and events of our best scientific theories can be *detected* with the aid of scientific instruments, but they cannot be *observed*. As we have seen in Chap. 3 (Sect. 3.3), constructive empiricists argue that when scientists use instruments, such as electron microscopes and radio telescopes, they are not *observing* theoretical entities, such as viruses and cosmic radiation. Rather, scientists are merely *detecting* these unobservables.

Accordingly, an antirealist like Bas van Fraassen (1980), could insist that Maxwell's Slippery Slope Argument for Instrumental Observation is unsound. In particular, van Fraassen would reject premise (P3) of Maxwell's Slippery Slope Argument for Instrumental Observation. For van Fraassen, the fact that no principled distinction can be drawn between *X* and *Y* does not necessarily mean that there is no useful distinction between *X* and *Y*. The distinction between observable entities and unobservable entities can still be useful, van Fraassen would argue, since there are clear-cut cases of observable entities, such as rocks and trees, and clear-cut cases of unobservable entities, such as electrons and viruses. If van Fraassen is right about this, then although Maxwell's Slippery Slope Argument for Instrumental Observation is valid, it cannot be said to be a sound argument.

## 4.3   The Argument from Observability in Principle

Grover Maxwell also tried to show that belief in the existence of the theoretical entities of our best scientific theories, which are unobservable for scientific antirealists because they cannot be directly observed with the naked eye, is justified by arguing that, *in principle*, any theoretical entity can become "observable." That is to say, if there are no *a priori* or non-empirical reasons to think that some entities, processes, and events are just unobservable, then, *in principle*, those entities, processes, and events can become observable. This argument can be found in Maxwell's seminal (1962, p. 11) paper.

> Maxwell, Grover. 1962. The ontological status of theoretical entities. In *Scientific Explanation, Space, and Time: Minnesota Studies in the Philosophy of Science*, eds., H. Feigl and G. Maxwell, 181–192. Minneapolis: University of Minnesota Press.

> *we are operating, here, under the assumption that it is theory, and thus science itself, which tells us what is or is not, in this sense, observable (the 'in principle' seems to have become superfluous). And this is the heart of the matter; for it follows that, at least for this sense of 'observable,' there are no a priori or philosophical criteria for separating the observable from unobservable.*

As I understand it, the argument that Maxwell is making in this passage can be stated in canonical (or standard) form as follows:

(P1) Science tells us what is and what is not observable.
(P2) If science tells us what is and what is not observable, then what is and what is not observable is determined empirically.

Therefore,

(C1) What is and what is not observable is determined empirically. [from (P1) & (P2)]

(P3) If what is and what is not observable is determined empirically, then there are no *a priori* or non-empirical reasons to think that some things are just unobservable in principle.

Therefore,

(C2) There are no *a priori* or non-empirical reasons to think that some things are just unobservable in principle. [from (C1) & (P3)]

(P4) If there are no *a priori* or non-empirical reasons to think that some things are just unobservable in principle, then belief in the existence of both observables and unobservables is justifiable in principle.

Therefore,

(C3) Belief in the existence of both observables and unobservables is justifiable in principle. [from (C2) & (P4)]

This line of reasoning is valid. That is to say, at each step, the premises provide logically conclusive support for the conclusion that follows from those premises. The next question, then, is whether the premises are actually true. Is Maxwell's Argument from Observability In Principle sound?

An antirealist like Bas van Fraassen would not accept the premises of this argument. In particular, van Fraassen would take issue with premise (P3) of Maxwell's Argument from Observability In Principle. As we have seen in Chap. 3 (see Sect. 3.3), van Fraassen would insist on the distinction between *detecting* and *observing*. More specifically, van Fraassen would argue that "the observability of interest is relativized to 'us'" (Monton and Mohler 2017). On this interpretation of 'observable', scientific and technological advancements could not make unobservable entities observable, no matter how sophisticated our scientific instruments become. Advances in science and technology can only make unobservable entities *detectable*. The only way in which an unobservable entity could become observable is if human sense organs somehow improve in a radical way, that is, in ways that would now be considered science fiction rather than science fact. For example, a scientific realist like Stathis Psillos (1999, p. 183) would argue that, in principle, we could "see a virus (without using a microscope, that is)" if "we had the technology to massively enlarge them, or to reduce some humans to so miniscule a size as to fit inside a minute capsule and be injected into someone's bloodstream." But this is science fiction, van Fraassen would protest. On the one hand, it is reasonable to think that advances in science and technology could get us closer to distant objects that currently we cannot observe without the aid of telescopes. For example, advances in space travel technology, such as the New Horizons space probe, might enable us to get a closer look at Pluto. This is because Pluto is a massive celestial object. If we were standing next to it, we would be able to see it. On the other hand, we have no reason to expect that advances in science and technology could shrink us to the size of a virus. Such things only happen in the movies, for example, *Honey, I Shrunk the Kids* (1989), hence, they are science fiction, not science fact. Unlike Pluto, even if we were standing next to a virus, we would still not be able to see it.

If van Fraassen is right about this, then, even if what is observable is determined by science facts, that is, empirically, this does not necessarily mean that there are no other reasons to think that some entities are unobservable in principle. Those reasons have to do with the limits of our human sense perception and the limits of our technology. For these reasons, although Maxwell's Argument from Observability In Principle is valid, it cannot be said to be a sound argument.

## 4.4   The Argument from Corroboration

As we have seen in Chap. 3 (see Sect. 3.4), Entity Realism is the view that the theoretical entities posited by our best scientific theories are real. One of the key arguments for this view is an argument from the manipulation of theoretical entities, which is based on Ian Hacking's (1983, p. 23) famous slogan for Entity Realism, "if you can spray them, they are real." That is to say, if we can do things with the theoretical entities posited by scientific theories, such as shooting electrons through a double-slit apparatus using an electron gun or combining genetic material from various sources to create recombinant DNA using methods of genetic recombination such as molecular cloning, then we have a good reason to believe that these theoretical entities (for example, electrons and DNA molecules) are real.

Another key argument for realism about the theoretical entities of our best scientific theories proceeds from corroboration rather than manipulation. That is to say, we have a good reason to believe that a theoretical entity is real, the argument goes, when results obtained by distinct experimental means point to the existence of that theoretical entity. A version of this argument can be found in Hacking's seminal (1983, p. 201) book.

Hacking, Ian. 1983. *Representing and intervening: introductory topics in the philosophy of natural science*. New York: Cambridge University Press.

*Two physical processes – electron transmission and fluorescent re-emission – are used to detect the bodies. These processes have virtually nothing in common between them. They are essentially unrelated chunks of physics. It would be a preposterous coincidence if, time and again, two completely different physical processes produced identical visual configurations which were, however, artifacts of the physical processes rather than real structures in the cell.*

As I understand it, the argument that Hacking is making in this passage can be stated in canonical (or standard) form as follows:

(P1) Different and independent physical means of detection point to the existence of the same theoretical entity *x*.
(P2) The best explanation for the phenomenon described in (P1) is that *x* is real.
(P3) No other explanation explains the phenomenon described in (P1) as well as (entity) realism about *x* does.

Therefore,

(C) *x* is real.

Like the Positive Argument for scientific realism (see Sect. 4.1), the Argument from Corroboration for Entity Realism is an instance of Inference to the Best Explanation

(IBE). As we have seen in Sect. 4.1, IBE is typically construed as an ampliative (or non-deductive) form of argumentation that proceeds from a phenomenon that requires an explanation to the conclusion that the best explanation for that phenomenon is probably true. In the case of the Argument from Corroboration, the phenomenon that demands an explanation is the fact that one and the same theoretical entity is revealed by distinct physical modes of experimentation or detection. This corroboration of evidence for the existence of a theoretical entity is supposed to be explained by the hypothesis that the theoretical entity is real, that is, by Entity Realism. If Entity Realism does indeed provide the best explanation for the fact that detections of theoretical entities can be corroborated by different physical means, then that would be a good reason to believe that Entity Realism is probably correct. In other words, the premises of the Argument from Corroboration, if true, would provide strong probable (though not conclusive) support for its conclusion. Given that the Argument from Corroboration for Entity Realism is not meant to be a deductive argument, the next question is not whether it is valid or invalid, but rather whether it is strong or weak.

As we have already seen in Sect. 4.1, some antirealists take issue with IBE. According to Bas van Fraassen (1980), from the fact that an explanation is the best one we could come up with, it does not follow that the explanation is probably true. For we might be working with a bad lot of explanations and the likely true explanation simply did not occur to us. If van Fraassen is right about this, then any inference to the best explanation would be a weak argument. For, even if no other explanation explains $P$ (for example, the corroboration of evidence for the existence of the same theoretical entity $x$ by independent physical means of detection) as well as $E$ (for example, Entity Realism) does, $E$ would still not be more probable or likely to be true if $E$ is simply the best of a "bad lot" of explanations.[15]

For the sake of argument, however, let us grant that IBE is a legitimate form of non-deductive (or inductive) argumentation. The next question, then, is whether the premises of the Argument from Corroboration for Entity Realism are in fact true? Is the Argument from Corroboration for Entity Realism a cogent argument? Premise (P1) is not in question, of course, for it states the phenomenon to be explained. So, to determine the cogency of the Argument from Corroboration we need to ask the following question: Does Entity Realism provide the best explanation for the phenomenon of corroborating evidence? Or are there alternative explanations that explain this phenomenon just as well as Entity Realism does? Antirealists may have such an alternative explanation in the form of what Bas van Fraassen calls "public hallucinations." According to van Fraassen (2001, p. 156), "Nature creates public hallucinations," for example, rainbows. Rainbows are *hallucinations* insofar as "there is no real material shining arch standing above the earth, although at first it looks that way" (van Fraassen 2001, p. 156), but also *public* insofar as "I see a rainbow and you say you see it too." Similarly, van Fraassen argues, images, shadows,

---

[15] We will revisit van Fraassen's criticism against IBE as inference from a "bad lot" of explanations in Chap. 5, Sect. 5.2, and Chap. 6, Sect. 6.4.

reflections, and other things detected by scientific instruments are public hallucinations. If van Fraassen is right about this, then there is no need to posit the existence of a real entity when different and independent physical means of detection point to it just as there is no reason to posit the existence of a real material shining arch standing above the Earth when you and I both see a rainbow. Like the rainbow, the entity our independent physical means of detection allegedly point to is a public hallucination. In other words, even though I see a rainbow and you see a rainbow, too, so we have different physical means of detection (that is, my eyes and your eyes) that point to the existence of the same entity (namely, a rainbow), we would *not* infer from this that there is a real material shining arch standing above the Earth because there is no need to postulate the existence of a real material shining arch standing above the Earth. A much better explanation is that the rainbow is a public hallucination.

Similarly, even if different physical means of detection point to the existence of the same theoretical entity $x$, we should *not* infer from this that $x$ is real because there is no need to postulate the existence of a real $x$. A much better explanation is that $x$ is a public hallucination. If this is correct, then (P3) of the Argument from Corroboration cannot be said to be true as antirealists would argue that there are alternative explanations for the phenomenon of corroboration that can explain it just as well as Entity Realism can. And if there are alternative explanations for the phenomenon of corroboration that can explain it just as well as Entity Realism can, then the Argument from Corroboration, even if strong, cannot be said to be a cogent argument.

More generally, while the physical means of detecting a theoretical entity may be theoretically independent (for example, detection by optical microscopy, digital microscopy, or electron microscopy), they are still *theoretically laden* means of detection. That is to say, these physical means of detection are based on theoretical knowledge from physics, optics, and other scientific fields of study. But this is precisely the sort of theoretical knowledge that antirealists deny. As we have seen in Chap. 2, antirealists generally think that science cannot (and does not) yield theoretical knowledge (that is, knowledge of unobservables). Insofar as the physical means used to detect theoretical entities (for example, electron transmission and fluorescent re-emission) are dependent on theoretical knowledge, and antirealists think that such theoretical knowledge is impossible to have in science, antirealists are unlikely to be impressed by the Argument from Corroboration. As Michela Massimi (2004, p. 39) puts it, "evidence in favour of [unobservable entities, such as] colored quarks [comes] from extraordinarily theory-loaded experiments." One cannot be committed to the existence of the unobservable entities, such as colored quarks, without thereby being committed to the associated theories, such as Quantum Chromodynamics (QCD). Given that antirealists tend to be agnostic about scientific theories like QCD, they would remain agnostic about the existence of the associated unobservable entities as well, the Argument from Corroboration notwithstanding. For these reasons, then, the Argument from Corroboration, even if strong, cannot be said to be a cogent argument.

## 4.5   The Argument from the Exponential Growth of Science

According to Ludwig Fahrbach (2011, p. 148), "both the number of journal articles and the number of scientists have grown with a doubling rate of 15–20 years." In other words, science is growing at an exponential rate. This exponential growth of science means that "at least 95% of all scientific work ever done has been done since 1915, and at least 80% of all scientific work ever done has been done since 1950" (Fahrbach 2011, p. 148). When we take into consideration the exponential growth of science, Fahrbach argues, we have a good reason to believe that the best scientific theories will remain stable over time, contrary to the antirealist "Graveyard" Argument (also known as the "pessimistic induction," see Chap. 5, Sect. 5.1). Fahrbach makes this argument in his (2011, p. 153) paper.

> Fahrbach, Ludwig. 2011. How the growth of science ends theory change. *Synthese* 180 (2): 139–155.

> *the fact that our current best theories have not been empirically refuted, but have been entirely stable for most of the history of science (weighted exponentially) invites an optimistic meta-induction to the effect that they will remain stable in the future, i.e., all their empirical consequences which scientists will ever have occasion to compare with results from observation at any time in the future are true.*[16]

As I understand it, the argument that Fahrbach is making in this passage can be stated in canonical (or standard) form as follows:

> (P) Our current best scientific theories have been entirely stable for most of the history of science (weighted exponentially).
>
> Therefore,
>
> (C) Our current best scientific theories will remain stable in the future.

This Argument from the Exponential Growth of Science is supposed to be an inductive, not a deductive, argument. So, the next question is not whether this argument is valid or invalid, but rather whether it is strong or weak. Does the phenomenon of the exponential growth of science provide strong inductive support for the conclusion that our current best theories will not be discarded in the future? Well, it depends on exactly how many scientific theories have been entirely stable for most of the history of science. In Mizrahi (2013), I compared random samples of theoretical posits and found that there are many more currently accepted (or stable) theoretical posits than abandoned (or discarded) theoretical posits. The ratio is approximately 3 : 10; that is, for every three abandoned theoretical posits, there are roughly ten stable theoretical posits that are currently accepted by practicing scientists (see also Mizrahi 2016a). I do not use these data to make an argument for scientific realism (for reasons that will become clear in Chap. 6). Instead, I take this empirical evidence from random samples to amount to some negative evidence

---

[16] Robert Nola (2008) also argues that the history of science provides evidence for an optimistic, rather than a pessimistic (see Chap. 5, Sect. 5.1), induction that supports scientific realism. Cf. Mizrahi (2013).

against "the historical graveyard of science" (Frost-Arnold 2011, p. 1138) picture (see Chap. 5, Sect. 5.1) in the following way (Mizrahi 2016a, p. 267):

1. If the history of science were a graveyard of dead theories and abandoned posits, then random samples of scientific theories and theoretical posits would contain significantly more dead theories and abandoned posits than live theories and accepted posits.
2. It is not the case that random samples of scientific theories and theoretical posits contain significantly more dead theories and abandoned posits than live theories and accepted posits.
3. Therefore, it is not the case that the history of science is a graveyard of dead theories and abandoned posits.

Nevertheless, some scientific realists might be inclined to use such empirical data as positive evidence for (P), that is, that there are many more stable than discarded scientific theories, which would then provide the inductive basis for an inductive inference to the conclusion that our current best scientific theories will probably remain stable as well. If this is correct, then the Argument from the Exponential Growth of Science can be said to be a strong argument.

If the Argument from the Exponential Growth of Science is a strong argument, the next question is whether its premise is true. If its premise is true, then the Argument from the Exponential Growth of Science can be said to be a cogent argument. Antirealists might object to (P) by claiming that it unwarrantedly assumes a privileged perspective for current scientists. As Brad Wray (2018, pp. 94–95) puts it:

> Consider the scientists and philosophers of science who lived and worked between 1890 and 1950. These scientists could also argue (then) that they were responsible for 80% of the scientific research ever produced. And they too would likely note that few of the theories developed during their era had been thrown out, at least by 1950. Indeed, today we see the science of their era differently, but they did not have the perspective that we have now.

If Wray is right about this, then we cannot say with sufficient confidence that our current best theories have been entirely stable for most of the history of science (weighted exponentially), that is, that (P) is true, because we may be in the position of early twentieth century scientists who thought that their best theories were stable. According to Wray (2018, p. 95), "we must take seriously the fact that once successful theories are often discarded later, on the grounds that they are false" (see Chap. 5, Sect. 5.1).

Scientific realists could respond to this antirealist objection to (P) by claiming that early twentieth century scientists did not actually witness the exponential growth of science that Fahrbach is referring to in his Argument from the Exponential Growth of Science. Recall that, according to Fahrbach (2011, p. 148), "at least 95% of all scientific work ever done has been done since 1915, and at least 80% of all scientific work ever done has been done *since 1950*" (emphasis added). This means that Wray's early twentieth century scientists, that is, "scientists and philosophers of science who lived and worked between 1890 and 1950" (Wray 2018, p. 95), did not witness the sort of doubling rates in scientists and scientific journal articles that Fahrbach calls "the exponential growth of science."

Another problem with Wray's antirealist objection against (P) is that it seems to presuppose "that once successful theories are often discarded later, *on the grounds that they are false*" (Wray 2018, p. 95; emphasis added). If antirealists were to concede that we could know that a scientific theory is false, it seems that they would also have to concede that theoretical knowledge is possible in science after all. For, if we can know that scientific theory $T$ is false, then we can also know that '$T$ is false' is true. Either way, we would have theoretical knowledge in science. Of course, antirealists would not want to concede that we could have theoretical knowledge in science, as we have seen in Chap. 2. So, instead, just as scientific realists take the stability of a theory as a sign (though not a sure sign) that it is approximately true, antirealists must take the abandonment of a theory as a sign (though not a sure sign) that it is probably false (see Chap. 5, Sect. 5.1).

Be that as it may, Wray's objection against the premise of the Argument from the Exponential Growth of Science points to a fundamental disagreement between scientific realists and antirealists, which characterizes much of the scientific realism/ antirealism debate in contemporary philosophy of science. Antirealists or "Pessimists look at the historical record of science and see failure, whereas realists look at the same historical record and see success" (Mizrahi 2013, p. 3214). This might lead one to think that the historical record of science is indeterminate between scientific realism and antirealism. On the one hand, scientific realists can select historical facts and case studies from the history of science that support a realist position about science. On the other hand, antirealists can select historical facts and case studies from the history of science that support an antirealist position about science. If that is the case, then historical facts and case histories do not favor a realist over an antirealist position, and vice versa. This seems to be the case with respect to the historical evidence used by Kyle Stanford (2006) to advance an antirealist argument (see Chap. 5, Sect. 5.4). That is:

> the mere fact that one can draw both pessimistic [or antirealist] and optimistic [or realist] conclusions from the historical record [of science] shows that [...] historical evidence is indeterminate between a pessimistic [or antirealist] interpretation and an optimistic [or realist] interpretation [of the historical record of science] (Mizrahi 2015, p. 145).

In that respect, recall the discussion of case studies from Chap. 2 (see Sect. 2.2). As we have seen in Chap. 2, scientific realists and antirealists use case studies from the history of science to argue for their respective positions. Sometimes they even use the same case study to argue for scientific realism or for antirealism. Such is the case with the phlogiston case study. Antirealists have used it to argue against scientific realism (of some variety or another), whereas scientific realists have used it to argue for scientific realism (of some variety or another). However, if the same evidence can be used to argue for opposing positions, then that evidence does not favor one of those positions over the other.

Indeed, antirealists used to take this point as a premise in an argument against scientific realism. This argument is known as the argument from the underdetermination of theories by evidence. Although this argument is no longer widely discussed in the contemporary literature on the scientific realism/antirealism debate, it is useful to discuss it briefly here. According to Stathis Psillos (1999, p. 162), the argument from the underdetermination of theories by evidence goes like this:

> two theories which are observationally indistinguishable, i.e. they entail exactly the same observational consequences, are *epistemically indistinguishable*, too, being equally well supported by the evidence. Hence, the argument concludes, *there are no positive reasons to believe in one rather than the other* (emphasis added).[17]

Now, let us apply this argument to scientific realism and antirealism. We have two competing theories, for example, Explanationist Realism and Constructive Empiricism, which are observationally indistinguishable, given that they have the same observational consequences concerning scientific practice. More explicitly, both Explanationist Realism and Constructive Empiricism predict the empirical (that is, explanatory and predictive) success of science as both are supposed to explain that success. In addition, both Explanationist Realism and Constructive Empiricism appear to be equally well supported by the historical evidence, too, as both scientific realists and antirealists use case studies from the history of science – even the same case studies, such as the phlogiston case – as supporting historical evidence for Explanationist Realism and Constructive Empiricism, respectively. This means that Explanationist Realism and Constructive Empiricism are not only observationally indistinguishable but also epistemically indistinguishable, at least insofar as historical evidence is concerned. Therefore, it follows that there are no positive *historical* reasons to believe in one rather than the other. In other words, both scientific realism and antirealism are underdetermined by the historical evidence (Mizrahi 2016a).

Since the Argument from the Exponential Growth of Science relies on historical facts about science (specifically, what Fahrbach calls "the exponential growth of science"), and such historical evidence is indeterminate between realist and antirealist positions in the scientific realism/antirealism debate, it follows that historical facts about science do not favor scientific realism over antirealism or antirealism over realism. That is to say, as philosophical theories about science, scientific realism and antirealism are observationally indistinguishable because they imply the same observational consequences about the history of science, as evidenced by the fact that scientific realists and antirealists often use historical evidence (even the same historical evidence, such as the phlogiston case study) to support scientific realism or antirealism, respectively. (See Chap. 2, Sect. 2.2). By the argument from the underdetermination of theories by evidence, then, scientific realism and antirealism are epistemically indistinguishable, given that they are equally well supported by the historical evidence. Therefore, there are no positive *historical* reasons to believe in scientific realism rather than antirealism. For this reason, the Argument

---

[17]Thomas Bonk (2008) provides a book-length treatment of the argument from the underdetermination of theories by evidence. See also Okasha (2002) and Ivanova (2010).

from the Exponential Growth of Science, which is an argument from historical facts about science (specifically, what Fahrbach calls "the exponential growth of science"), even if strong, cannot be said to be a cogent argument for scientific realism.

## 4.6  Summary

In the contemporary scientific realism/antirealism debate, the positive argument for scientific realism that has attracted the most attention from scientific realists and antirealists alike is the so-called "miracle" or "no miracles" argument. According to this argument, our best scientific theories are approximately true because that is the best explanation for their empirical success. Antirealists object to this argument by rejecting Inference to the Best Explanation (IBE) entirely as an illegitimate form of inference and by offering alternative explanations for the empirical success of our best scientific theories (see Sect. 4.1). Scientific realists attempt to undermine the antirealist distinction between observables and unobservables by means of two arguments. The first is a slippery slope argument that purports to show that the observable/unobservable distinction is untenable. That is to say, there is no principled difference between observing without instruments and observing with instruments (see Sect. 4.2). The second is an argument that purports to show that anything can be observable in principle. That is to say, there are no *a priori* (or non-empirical) reasons to think that some things are unobservable (see Sect. 4.3). Antirealists object to these arguments by drawing a distinction between observation and detection. Entity realists argue that corroborating evidence, that is, evidence obtained by distinct physical means of detection, for the existence of a theoretical entity is best explained by the hypothesis that the corroborated theoretical entity is real. Antirealists object to this argument by pointing out that the physical means of detecting theoretical entities are theory-laden, that is to say, their operation depends on the very theoretical knowledge that antirealists argue is impossible to obtain in science (see Sect. 4.4). Finally, some scientific realists argue that, when we take into consideration the exponential growth of science, we have a good reason to believe that the best scientific theories will remain stable over time (see Sect. 4.5). Antirealists, on the other hand, tend to be pessimistic about the prospects of our best scientific theories.

## Glossary

**Antirealism** An agnostic or skeptical attitude toward the theoretical posits (that is, unobservables) of scientific theories. Antirealism comes in different varieties, such as Constructive Empiricism (see Chap. 3, Sect. 3.3) and Instrumentalism (see Chap. 3, Sect. 3.2).

**Approximate truth** Closeness to the truth or truthlikeness. To say that a theory is approximately true is to say that it is close to the truth. According to some scientific realists, approximate truth is the aim of science. (See Chap. 2, Sect. 2.1).

**Circular definition** A definition is (viciously) circular when the term to be defined, or some variation thereof, is used in the definition itself. (See Sect. 4.1).

**Circular reasoning** An argument is (viciously) circular when its conclusion appears as one of the premises in the argument. Also known as begging the question. (See Sect. 4.1).

**Direct observation** Observation with the naked eye, without the use of scientific instruments, such as microscopes and telescopes, as opposed to instrument-aided observation. (See Chap. 3, Sect. 3.3).

**Empirical success** A scientific theory is said to be empirically successful just in case it is both explanatorily successful (that is, it explains natural phenomena that would otherwise be mysterious to us) and predictively successful (that is, it makes predictions that are borne out by observation and experimentation). (See Chap. 3, Sect. 3.1).

**Explanatory success** A scientific theory is said to be explanatorily successful just in case it explains natural phenomena that would otherwise be mysterious to us. (See Chap. 3, Sect. 3.1).

**The exponential growth of science** The claim that scientific output grows at an exponential rate, with at least 95% of all scientific work having been done since 1915, and at least 80% of all scientific work having been done since 1950. (See Sect. 4.5).

**Fallacious argument** An argument whose premises fail to provide either conclusive or probable support for its conclusion (see also *invalid argument* and *weak argument*). (See Chap. 2, Sect. 2.2).

**Inference to the Best Explanation (IBE)** An ampliative (or non-deductive) form of argumentation that proceeds from a phenomenon that requires an explanation to the conclusion that the best explanation for that phenomenon is probably true. (See Sect. 4.1).

**Instrument-aided observation** Observation by means of scientific instruments, such as microscopes and telescopes, as opposed to direct or naked-eye observation. (See Chap. 3, Sect. 3.3).

**Invalid argument** A deductive argument in which the premises purport but fail to provide logically conclusive support for the conclusion. (See Chap. 1, Sect. 1.1).

**Modus ponens** A form of argument with a conditional premise, a premise that asserts the antecedent of the conditional premise, and a conclusion that asserts the consequent of the conditional premise. That is, "if $A$, then $B$, $A$; therefore, $B$," where $A$ and $B$ stand for statements. *Modus ponens* is a valid form of inference, and so an argument in natural language that takes this logical form is valid. On the other hand, the following logical form is invalid: "if $A$, then $B$, $B$; therefore, $A$." It is known as the fallacy of affirming the consequent. (See Sect. 4.1).

**Modus tollens** A form of argument with a conditional premise, a premise that denies the consequent of the conditional premise, and a conclusion that denies the antecedent of the conditional premise. That is, "if $A$, then $B$, not $B$; therefore, not *A*," where *A* and *B* stand for statements. *Modus tollens* is a valid form of inference, and so an argument in natural language that takes this logical form is valid. On the other hand, the following logical form is invalid: "if *A*, then *B*, not *A*; therefore, not *B*." It is known as the fallacy of denying the antecedent. (See Chap. 5, Sect. 5.1).

**Predictive success** A scientific theory is said to be predictively successful just in case it makes predictions that are borne out by observation and experimentation. (See Chap. 3, Sect. 3.1).

**Scientific realism** An epistemically positive attitude toward those aspects of scientific theories that are worthy of belief. Scientific realism comes in different varieties, such as Explanationist Realism (see Chap. 3, Sect. 3.1), Entity Realism (see Chap. 3, Sect. 3.4), Structural Realism (see Chap. 3, Sect. 3.5), and Relative Realism (see Chap. 6, Sect. 6.1).

**Selectionist explanation for the success of science** An explanation for the empirical success of our best scientific theories according to which a selection process akin to natural selection by which fit species survive and unfit species go extinct explains the survival of successful theories and the extinction of unsuccessful theories. (See Sect. 4.1).

**Theoretical virtues** Properties of scientific theories, such as unification, testability, coherence, and simplicity, that make theories that have them good theories. Scientific realists tend to think of such properties as epistemic or truth conducive, whereas antirealists tend to think of them as merely pragmatic. (See Sect. 4.1).

**Underdetermination of theories by evidence** An antirealist argument according to which, if two theories are observationally indistinguishable, then they are epistemically indistinguishable, and thus there are no positive reasons to believe in one over the other. (See Sect. 4.5).

**Weak argument** A non-deductive (or inductive) argument in which the premises purport but fail to provide probable support for the conclusion. (See Chap. 1, Sect. 1.1).

# References and Further Readings

Baker, A. (2010). Inference to the best explanation. In F. Russo & J. Williamson (Eds.), *Key terms in logic* (pp. 37–38). London: Continuum.

Bird, A. (2007). Inference to the only explanation. *Philosophy and Phenomenological Research, 74*(2), 424–432.

Bonk, T. (2008). *Underdetermination: An essay on evidence and the limits of natural knowledge.* Dordrecht: Springer.

Braithwaite, R. B. (1953). *Scientific explanation: A study of the function of theory, probability and law in science.* New York: Cambridge University Press.

Campos, D. G. (2011). On the distinction between Peirce's abduction and Lipton's inference to the best explanation. *Synthese, 180*(3), 419–442.

Fahrbach, L. (2011). How the growth of science ends theory change. *Synthese, 180*(2), 139–155.

Fine, A. (1986). Unnatural attitudes: Realist and instrumentalist attachments to science. *Mind, 95*(378), 149–179.

Fine, A. (1991). Piecemeal realism. *Philosophical Studies, 61*(1), 79–96.

Frost-Arnold, G. (2011). From the pessimistic induction to semantic antirealism. *Philosophy of Science, 78*(5), 1131–1142.

Hacking, I. (1983). *Representing and intervening: Introductory topics in the philosophy of natural science*. New York: Cambridge University Press.

Harman, G. H. (1965). The inference to the best explanation. *The Philosophical Review, 74*(1), 88–95.

Hempel, C. G. (1966). *Philosophy of natural science*. Englewood Cliffs: Prentice-Hall.

Ivanova, M. (2010). Pierre Duhem's good sense as a guide to theory choice. *Studies in History and Philosophy of Science Part A, 41*(1), 58–64.

Keas, M. N. (2018). Systematizing the theoretical virtues. *Synthese, 195*(6), 2761–2793.

Ladyman, J. (2002). *Understanding philosophy of science*. London: Routledge.

Ladyman, J. (2007). Ontological, epistemological, and methodological positions. In T. Kuipers (Ed.), *General philosophy of science: Focal issues* (pp. 303–376). Amsterdam: Elsevier.

Ladyman, J., Ross, D., Spurrett, D., & Collier, J. (2007). *Every thing must go: Metaphysics naturalized*. New York: Oxford University Press.

Laudan, L. (1981). A confutation of convergent realism. *Philosophy of Science, 48*(1), 19–49.

Massimi, M. (2004). Non-defensible middle ground for experimental realism: Why we are justified to believe in colored quarks. *Philosophy of Science, 71*(1), 36–60.

Maxwell, G. (1962). The ontological status of theoretical entities. In H. Feigl & G. Maxwell (Eds.), *Scientific explanation, space, and time: Minnesota studies in the philosophy of science* (pp. 181–192). Minneapolis: University of Minnesota Press.

Mizrahi, M. (2012). Why the ultimate argument for scientific realism ultimately fails. *Studies in History and Philosophy of Science Part A, 43*(1), 132–138.

Mizrahi, M. (2013). The pessimistic induction: A bad argument gone too far. *Synthese, 190*(15), 3209–3226.

Mizrahi, M. (2015). Historical inductions: New cherries, same old cherry-picking. *International Studies in the Philosophy of Science, 29*(2), 129–148.

Mizrahi, M. (2016a). The history of science as a graveyard of theories: A philosophers' myth? *International Studies in the Philosophy of Science, 30*(3), 263–278.

Mizrahi, M. (2016b). Historical inductions, unconceived alternatives, and unconceived objections. *Journal for General Philosophy of Science, 47*(1), 59–68.

Mizrahi, M. (2018). The "positive argument" for constructive empiricism and inference to the best explanation. *Journal for General Philosophy of Science, 3*, 1–6.

Mizrahi, M. (2020). Theoretical virtues in scientific practice: An empirical study. *The British Journal for the Philosophy of Science*.

Monton, B., Mohler, C. (2017). Constructive empiricism. In Zalta, E. N. (Ed.), *The Stanford encyclopedia of philosophy* (Summer 2017 ed.). https://plato.stanford.edu/archives/sum2017/entries/constructive-empiricism

Musgrave, A. (1988). The ultimate argument for scientific realism. In R. Nola (Ed.), *Relativism and realism in science* (pp. 229–252). Dordrecht: Kluwer Academic Publishers.

Nola, R. (2008). The optimistic meta-induction and ontological continuity: The case of the electron. In L. Soler, H. Sankey, & P. Hoyningen-Huene (Eds.), *Rethinking scientific change and theory comparison* (pp. 159–202). Dordrecht: Springer.

Okasha, S. (2002). Underdetermination, holism, and the theory/data distinction. *The Philosophical Quarterly, 52*(208), 303–319.

Psillos, S. (1999). *Scientific realism: How science tracks truth*. London: Routledge.

Psillos, S. (2006). Thinking about the ultimate argument for scientific realism. In C. Cheyne & J. Worrall (Eds.), *Rationality and reality: Conversations with Alan Musgrave* (pp. 133–156). Dordrecht: Springer.

Psillos, S. (2007). The fine structure of inference to the best explanation. *Philosophy and Phenomenological Research, 74*(2), 441–448.

Psillos, S. (2011). The scope and limits of the no miracles argument. In D. Dieks, W. J. Gonzalez, S. Hartmann, T. Uebel, & M. Weber (Eds.), *Explanation, prediction, and confirmation* (pp. 23–35). Dordrecht: Springer.

Putnam, H. (1975). *Mathematics, matter and method.* New York: Cambridge University Press.

Schindler, S. (2018). *Theoretical virtues in science: Uncovering reality through theory.* Cambridge: Cambridge University Press.

Stanford, K. P. (2000). An antirealist explanation of the success of science. *Philosophy of Science, 67*(2), 266–284.

Stanford, K. P. (2006). *Exceeding our grasp: Science, history, and the problem of unconceived alternatives.* New York: Oxford University Press.

van Fraassen, B. C. (2001). Constructive empiricism now. *Philosophical Studies, 106*(1–2), 151–170.

Wray, B. K. (2007). A selectionist explanation for the success and failures of science. *Erkenntnis, 67*(1), 81–89.

Wray, B. K. (2018). *Resisting scientific realism.* Cambridge: Cambridge University Press.

# Chapter 5
# Key Arguments Against Scientific Realism

**Abstract** In this chapter, I present in canonical (or standard) form and then evaluate key arguments against scientific realism (or for antirealism about science). The
first argument is known as the "pessimistic induction" or the "pessimistic meta-
induction." In its original formulation, attributed to Larry Laudan (Philos Sci
48(1):19–49, 1981), the argument is based on a list of theories that are supposed to
be counterexamples to the realist thesis that empirical success is a mark of (approximate) truth. Other formulations of the argument have it as an inductive argument
from a sample of theories that were discarded in the past to the conclusion that our
present scientific theories will probably be discarded as well. The second argument
is a positive argument for Bas van Fraassen's (The scientific image. Oxford
University Press, New York, 1980) antirealist position, namely, Constructive
Empiricism. The third argument proceeds from the observation that scientists cannot claim to be in a privileged epistemic position. According to this argument, it is
unlikely that our best scientific theories are (approximately) true because scientists
are not especially skilled at developing theories that are likely true or approximately
true. A contemporary proponent of this argument for antirealism is Brad Wray (Int
Stud Philos Sci 22(3):317–326, 2008). The fourth is an argument for antirealism
from what Kyle Stanford (Exceeding our grasp: science, history, and the problem of
unconceived alternatives. Oxford University Press, New York, 2006) calls the
"Problem of Unconceived Alternatives" (PUA). According to Stanford, the historical record of science reveals that past scientists typically failed to conceive of alternatives to their favorite, then-successful theories. This is supposed to make it more
likely that present scientists fail to conceive of alternatives to their favorite, now-
successful theories. For these reasons, Stanford argues, we should not believe our
present scientific theories are (approximately) true. The fifth argument purports to
show that many of the scientific theories we currently accept will be discarded
sometime in the future, as the "pessimistic induction" does, but it proceeds from the
premise that the research goals and interests of scientists change over time (Wray,
Brad K., Resisting scientific realism. Cambridge University Press, Cambridge, 2018).

**Keywords** Approximate truth · Cherry-picking · Constructive empiricism ·
Epistemic privilege · Explanationist realism · Hasty generalization · Historical
graveyard of science · Inference to the Best Explanation (IBE) · Predictive success ·
Problem of Unconceived Alternatives (PUA) · Random sampling · Representative
sample · Research goals · Research interests

© Springer Nature Switzerland AG 2020
M. Mizrahi, *The Relativity of Theory*, Synthese Library 431,
https://doi.org/10.1007/978-3-030-58047-6_5

In this chapter, I present in canonical (or standard) form and then evaluate key arguments against scientific realism (or for antirealism about science). The first argument is known as the "pessimistic induction" or the "pessimistic meta-induction." Even though "induction" is part of the name, there is no agreement among philosophers of science on whether to construe this argument as an inductive or a deductive argument. In its original formulation, attributed to Larry Laudan (1981), the argument is based on a list of theories that are supposed to be counterexamples to the realist thesis that empirical success is a mark of (approximate) truth. Laudan's (1981, p. 33) list includes scientific theories from past science, such as the humoral theory of medicine, the phlogiston theory of chemistry, the vital force theory of physiology, and theories of spontaneous generation, among others. The list includes twelve scientific theories in total, although Laudan (1981, p. 33) claims that the list "could be extended *ad nauseum.*" Other formulations of the argument have it as an inductive argument from a sample of theories that were discarded in the past to the conclusion that our present scientific theories will probably be discarded as well. The second argument is a positive argument for Bas van Fraassen's (1980) antirealist position, namely, Constructive Empiricism. The third argument proceeds from the observation that scientists cannot claim to be in a privileged epistemic position (Lipton 1993). According to this argument, it is unlikely that our best scientific theories are (approximately) true because it is not the case that scientists are especially skilled at developing theories that are likely true or approximately true. A contemporary proponent of this argument for antirealism is Brad Wray (2008) and (2018). The fourth is an argument for antirealism from what Kyle Stanford (2006) calls the "Problem of Unconceived Alternatives" (PUA). According to Stanford, the historical record of science reveals that past scientists typically failed to conceive of alternatives to their favorite, then-successful theories. This is supposed to make it more likely that present scientists fail to conceive of alternatives to their favorite, now-successful theories. For these reasons, Stanford argues, we should not believe our present scientific theories are (approximately) true. The fifth argument purports to show that many of the scientific theories we currently accept will be discarded sometime in the future, like the "pessimistic induction" does, but it proceeds from the premise that the research goals and interests of scientists change over time (Wray 2018).

## 5.1   The "Graveyard" Argument

If the Positive Argument for scientific realism, the so-called "no miracles" argument (see Chap. 4, Sect. 4.1), is the realist argument that has dominated the scientific realism/antirealism debate in contemporary philosophy of science, the so-called "pessimistic induction" is its antirealist counterpart. There is some debate among philosophers of science concerning the structure of this antirealist argument.[1] Some think that it is an inductive argument from a sample, others interpret it as a deductive argument, and still others understand it as an argument from counterexamples.[2] What all these different interpretations of the argument share in common is an assumption that the history of science is a "graveyard of dead epistemic objects" (Chang 2011, p. 426), that is, discarded theories and abandoned theoretical posits.[3] From this historical understanding of science as a "graveyard of dead epistemic objects" (Chang 2011, p. 426), antirealists then draw the conclusion that current (and future) scientific theories and theoretical posits will end up in that "graveyard" as well. That is why I refer to this argument as the "Graveyard" Argument. Still, antirealists get to the conclusion that our best current theories will end up in the "graveyard" of science in various ways. Some do it by inductive reasoning. An inductive version of the "Graveyard" Argument can be found in Brad Wray's (2018, pp. 68–69) book-length defense of antirealism.

Wray, Brad K. 2018. *Resisting scientific realism.* Cambridge: Cambridge University Press.

> *A Pessimistic Induction is an inductive argument that draws a conclusion from the rejection of many successful scientific theories in the past. Sometimes a conclusion is drawn about the prospects of the theories that are currently accepted, and sometimes an inference is drawn about the prospects of future theories, those not yet developed or entertained by scientists.*

As I understand it, the argument that Wray is making in this passage can be stated in canonical (or standard) form as follows:

(P) Many successful scientific theories were rejected in the past.
    Therefore,

(C) Current (and future) successful scientific theories will be rejected.[4]

---

[1] Arguably, this is partly because the original presentation of the argument that came to be known as the "pessimistic induction," namely, Laudan (1981), lends itself to different interpretations. For a detailed discussion, see Mizrahi (2013a).

[2] See, for example, James Ladyman's (2011) for a discussion of the "pessimistic induction" as an argument from counterexamples against scientific realism. Ladyman (2011) argues that Structural Realism (see Chap. 3, Sect. 3.5) can account for one of those counterexamples, namely, the phlogiston theory.

[3] Peter Lipton (2005, p. 1265) states the argument as follows: "The history of science is a graveyard of theories that were empirically successful for a time, but are now known to be false, and of theoretical entities—the crystalline spheres, phlogiston, caloric, the ether and their ilk—that we now know do not exist. Science does not have a good track record for truth, and this provides the basis for a simple empirical generalization. Put crudely, all past theories have turned out to be false, therefore it is probable that all present and future theories will be false as well. That is the pessimistic induction." In Mizrahi (2016a), I provide empirical evidence against the "graveyard" picture of the history of science. Cf. Tulodziecki (2017).

[4] Anjan Chakravartty (2008, p. 152) states the argument as follows: the pessimistic induction is "a two step worry. First, there is an assertion to the effect that the history of science contains an impressive graveyard of theories that were previously believed, but subsequently judged to be false

Since this argument is not meant to be a deductive argument, the next question is not whether this argument is valid or invalid, but rather whether it is strong or weak. Does the premise (P) successfully provide probable support for the conclusion (C)? Well, it depends on exactly how many successful scientific theories were rejected and from which periods of past science. Suppose that the population of past successful scientific theories consists of one hundred theories. How many of them would have to be rejected to justify the inference to the conclusion that current (and future) successful scientific theories will be rejected as well? At least fifty? More than fifty? Recall that Laudan's list of "previously successful but now dead" scientific theories contains twelve scientific theories. Throughout the history of science, however, hundreds of theories were proposed by scientists. Again, suppose that the population of past successful scientific theories consists of one hundred theories. This means that 12% of those theories, that is, twelve out of one hundred, are "previously successful but now dead" scientific theories. Is 12% of a sample enough to draw general conclusions about the general population, or make predictions about any given member of the general population, beyond the sample?

As we have seen in Chap. 1 (see Sect. 1.1), in inductive prediction, given that $X$ percent of sampled things, $F$s, have a particular property, $G$, we are entitled to conclude that, with a probability of $X$ percent, a new $F$ that has not been observed or surveyed yet will also have the property, $G$, provided that $X$ is greater than 50%, and there is no evidence that the new $F$ is unlike previously observed $F$s (Schurz 2019, p. 2). Accordingly, given that only 12% of past scientific theories "previously successful but now dead" scientific theories, as we are supposing for the sake of argument, we are justified in concluding, with a probability of 12%, that any given current scientific theory that is successful will end up dead, that is, will be abandoned by scientists in the future. But this means that any given current scientific theory that is successful has an 82% probability of being retained rather than abandoned. Therefore, for any given current scientific theory that is successful, it is much more likely that this theory will be retained rather than abandoned, given the evidence we have. For these reasons, some philosophers of science have argued that the sample of successful scientific theories that were rejected in the past is not large enough to warrant the inductive inference to (C). From a handful of examples of rejected theories, such as the phlogiston theory and the caloric theory of heat (Laudan 1981, p. 33), no strong inductive inferences can be drawn to the conclusion that current (and future) successful scientific theories will be rejected.[5]

In addition to being too small to warrant an inductive inference to (C), or any inductive predictions about future scientific theories, the sample of past scientific theories on which the pessimistic induction is supposed to be based might also be unrepresentative of science as a whole. For, as we have seen, the examples typically cited in support of (P) tend to be from ancient science (for example, the humoral

---

[...]. Second, there is an induction on the basis of this assertion, whose conclusion is that current theories are likely future occupants of the same graveyard." See also Ladyman (2007, p. 345).

[5] For more on this criticism of the pessimistic induction, see Park (2011), Fahrbach (2011), Mizrahi (2013a), and Mizrahi (2016a).

theory of illness), medieval science (for example, the crystalline spheres), and early modern science (for example, the phlogiston theory) (Laudan 1981, p. 33). In that respect, these examples of rejected theories come from the distant past of science. But most current scientific theories are not from the distant past of science (see Chap. 4, Sect. 4.5 on the "exponential growth of science"). This means that the examples of discarded theories that are typically cited in support of (P) are not representative of science as a whole.

Finally, the examples of discarded theories of past science that are typically cited in support of (P) are not only too few and unrepresentative to warrant an induction about science as a whole but also cherry-picked instead of randomly selected. A strong inductive argument from a sample must be based on a randomly selected sample. If the sample is cherry-picked, rather than randomly selected, then it cannot be representative of the target population as a whole (see Chap. 2, Sect. 2.2). For example, if I survey 2000 Republican voters and find that 60% of them are satisfied with the job that President Donald Trump is doing, I am not entitled to conclude from this that 60% of American voters are satisfied with the job that President Donald Trump is doing. For my sample of voters consists of Republican voters only, whereas the general population of voters contains Democrats and Independent voters as well. Similarly, if I survey scientific theories from the distant past only (that is, from the ancient, medieval, and early modern periods only) and find that 60% of them are no longer accepted by scientists, I am not entitled to conclude from this that 60% of current (and future) scientific theories will be rejected as well. For my sample of scientific theories consists of pre-eighteenth century theories only, whereas the general population of scientific theories contains theories from the eighteenth, nineteenth, and twentieth centuries as well. In fact, if "at least 95% of all scientific work ever done has been done since 1915, and at least 80% of all scientific work ever done has been done since 1950," as Ludwig Fahrbach (2011, p. 148) argues, then pre-twentieth century scientific theories cannot be representative of science as a whole (see Chap. 4, Sect. 4.5).

Accordingly, if the sample of past scientific theories used to support (P) is too small, non-randomly selected (that is, cherry-picked), and unrepresentative of scientific theories in general, then it cannot provide strong inductive support for the conclusion that current (and future) successful scientific theories will be rejected. Indeed, as we have seen in Chap. 2 (see Sect. 2.2), drawing general conclusions or making predictions from small, non-random, and unrepresentative samples is a mistake in reasoning called "hasty generalization." According to Patrick Hurley (2006, p. 131), "Hasty generalization is a fallacy that affects inductive reasoning. [...] The fallacy occurs when there is a reasonable likelihood that the sample is not representative of the group. Such a likelihood may arise if the sample is either too small or not randomly selected." Since the sample on which the inductive version of the "Graveyard" Argument is based is too small and non-random, it is unlikely to be a representative sample. For these reasons, the "Graveyard" Argument (or the so-called "pessimistic induction") cannot be said to be a strong inductive argument.[6]

Other antirealists get to the conclusion that, just like past scientific theories, current (and future) scientific theories will end up in the so-called "historical graveyard of science" (Frost-Arnold 2011, p. 1138) as well, not by inductive reasoning, but by deductive reasoning. A deductive version of the "Graveyard" Argument can be found in Timothy Lyons' (2017, p. 3209) paper, where he argues that "the historical argument [that is, the "Graveyard" Argument] involves neither a meta-induction nor a conclusion that our scientific theories are (likely) false." Lyons (2017, p. 3209) states his argument in canonical (or standard) form as follows:

1. If (a) the realist meta-hypothesis were true, then (b) none of the constituents genuinely deployed toward successes would be such that they cannot be approximately true.
2. However, (not-b) we do find constituents genuinely deployed toward success that cannot be approximately true.
3. Therefore, (not-a) the realist meta-hypothesis is false.

For Lyons, the premises of this argument purport to provide logically conclusive, as opposed to probable, support for the conclusion. In other words, he intends this argument to be a deductive argument (in particular, a *modus tollens*: If *A*, then *B*, not *B*; therefore, not *A*). Since the premises of Lyons' argument, namely, (1) and (2), successfully provide logically conclusive support for the conclusion, namely, (3), this argument can be said to be valid. The next question, then, is whether the premises are in fact true. Is Lyons' argument, which purports to be a deductive version of the "Graveyard" Argument, sound?

For Lyons (2017, p. 3204), "the realist meta-hypothesis" is the claim that "successful scientific theories are (approximately) true." If this realist meta-hypothesis were true, Lyons argues, then there would be no successful theories that are not approximately true. For, according to "the realist meta-hypothesis," *all* "successful scientific theories are (approximately) true," with no exception. If there are any exceptions, that is, any successful theories that are not approximately true, then "the realist meta-hypothesis" must be false, or so Lyons argues. Like other antirealists who believe that the history of science is a "graveyard of dead epistemic objects" (Chang 2011, p. 426), Lyons also believes that there are successful theories that are not approximately true. He thus concludes by deductive reasoning that "the realist meta-hypothesis," that is, the claim that all "successful scientific theories are (approximately) true," is false.

---

[6]According to Sherri Roush, the pessimist has to show not only that past scientific theories have been failures but also that past scientific methods are failed methods of scientific inquiry. "But even if we grant their unreliability [that is, the unreliability of past methods of scientific inquiry]," Roush (2010, p. 55) argues, "nothing follows from this about whether we have a right to our confidence in our particular theories unless" the pessimist can also show "that their unreliability is a reason to think we are unreliable," which has not been shown, "since the manifest difference in methods between us and our predecessors breaks the pessimist's induction."

Scientific realists, however, would not accept premise (1) of Lyons' argument because they typically do not endorse what Lyons (2017, p. 3204) calls "the realist meta-hypothesis," that is, the claim that all "successful scientific theories are (approximately) true." Rather, for contemporary scientific realists, after the so-called "selectivist turn" (see Chap. 2), predictive success is a *reliable indicator* of approximate truth, not a *sure sign* that a scientific theory must be approximately true. Saying that predictive success is a *reliable indicator* of approximate truth is compatible with there being some exceptions, that is, a few scientific theories that are predictively successful but not approximately true. To put it another way, success may be a reliable indicator of (approximate) truth, but this is compatible with some instances of successful theories that turn out not to be approximately true. Indeed, even before the so-called "selectivist turn," scientific realists were careful in the way they formulated what Lyons calls "the realist meta-hypothesis." For example, as he was advancing the Positive Argument for scientific realism (see Chap. 4, Sect. 4.1), Putnam (1975, p. 73) was careful to add that "terms in mature scientific theory *typically* refer (this formulation is due to Richard Boyd), that the theories accepted in a mature science are *typically* approximately true [...]—these statements are viewed by the scientific realist *not as necessary truth* but as part of the only scientific explanation of the success of science" (emphasis added). In other words, "successful scientific theories are (approximately) true" is not meant to be a necessary truth. If a theory is predictively successful, then that is a reason to believe that it is approximately true, but it is not a conclusive reason to believe that the theory is approximately true (Mizrahi 2013a, p. 3224).[7]

Conversely, then, a successful theory that is not approximately true does not constitute a conclusive reason to believe that predictive success and approximate truth are not reliably connected with each other such that predictive success is a *reliable indicator* of approximate truth. This is much like how an academically successful student who has a poor attendance record is not a conclusive proof that class attendance and academic success are not reliably connected with each other such that class attendance is a *reliable indicator* (or predictor) of academic success. This is because, on average, students with poor attendance records struggle to succeed academically. The fact that there are a few students who manage to succeed academically despite having a poor attendance record does not change the fact that class attendance is a reliable indicator (or predictor) of academic success. Accordingly, in much the same way that class attendance is a reliable indicator (or predictor) of academic success despite a few outliers (that is, some students who are academically successful despite having poor attendance records), realists would argue, predictive success is a reliable indicator (or predictor) of approximate truth despite a few outliers (that is, some scientific theories that are predictively successful but not even approximately true). For this reason, premise (1) of Lyons' argument may be false, or at least unacceptable to scientific realists, which means that this deductive formulation of the "Graveyard" Argument, albeit valid, cannot be said to be sound.

---

[7] See also Park (2019, p. 607).

Finally, some antirealists prefer to understand the "Graveyard" Argument as an argument from counterexamples. After all, if the history of science really is a "graveyard of dead epistemic objects" (Chang 2011, p. 426), as antirealists claim it is, then it should be fairly easy to find examples of scientific theories that were predictively successful but were later abandoned. Those examples would then serve as counterexamples against the realist claim that predictive success is a reliable indicator of approximate truth. As Peter Vickers (2019, p. 572) puts it:

> Putnam's statement [namely, that scientific realism "is the only philosophy that doesn't make the success of science a miracle" (Putnam 1975, p. 73; see Chapter 4, Section 4.1)] was immediately objectionable due to fairly straight forward historical 'counterexamples'. Most famously, of course, Laudan explicitly targeted Putnam in his historical 'confutation'. *For a counterexample to Putnam's success-to-truth inference all we need are examples where we have (significant) success, and yet what we infer certainly isn't true.* As Laudan argues (1981, pp. 24–26), Putnam is happy to accept that, for example, 'aether' is a non-referring term, and thus theories employing this concept are not even approximately true. Laudan then argues that nineteenth century aether theories were successful theories (pp. 26–27). And he goes on to give his famous list of other possible/probable *counterexamples* (p. 33) (emphasis added).

Understood as an argument from counterexamples, then, the "Graveyard" Argument takes the same form as Lyon's deductive version of the "Graveyard" Argument, that is, *modus tollens*: If $A$, then $B$, not $B$; therefore, not $A$. This argument can be stated in canonical (or standard) form as follows:

1. If all successful scientific theories are approximately true, then any successful scientific theory, $T$, must also be approximately true.
2. But a successful scientific theory, $T$, is not approximately true.
3. Therefore, it is not the case that all successful scientific theories are approximately true.

Since the premises of this argument, namely, (1) and (2), successfully provide logically conclusive support for the conclusion, namely, (3), this argument can be said to be valid. The next question, then, is whether the premises are in fact true. Is this argument from counterexamples, which purports to be a deductive version of the "Graveyard" Argument, sound?

As we have seen, Larry Laudan's (1981, p. 33) list of past scientific theories that were predictively successful but were later discarded includes the ancient humoral theory of illness, the crystalline spheres of medieval astronomy, and the phlogiston theory of early modern chemistry. Accordingly, one could take any of the "previously successful but now dead" scientific theories on Laudan's (1981, p. 33) list and plug it in as a value for the $T$ variable in the argument from counterexamples. This would make premise (2) of the argument from counterexamples true. But what about premise (1)? Are these and other "previously successful but now dead" theories of past science effective counterexamples against the realist thesis that predictive success is a reliable indicator of approximate truth?

Again, scientific realists would take issue with premise (1) of the argument from counterexamples, specifically with the claim that all successful scientific theories are approximately true. Counterexamples are effective only if their intended targets are construed as universal statements or generalizations. For example, a bird that cannot fly would count as an effective counterexample against the universal generalization "All birds can fly." On the other hand, a bird that cannot fly would not count as an effective counterexample against the claim that a keel is a reliable indicator of flying ability in birds. For the claim that birds with a keel on their breastbone are more likely to be flying birds than flightless birds is consistent with there being some flightless birds, such as ostriches and kiwis, with a keel on their breastbone (though greatly reduced). Likewise, a predictively successful but dead theory would count as an effective counterexample against the universal generalization "All successful theories are approximately true." As we have already seen, however, that is not the scientific realist's claim. The scientific realist's claim is that predictive success is a reliable indicator of approximate truth. Therefore, a predictively successful but dead theory does not count as an effective counterexample against the claim that predictive success is a reliable indicator of approximate truth. For the claim that predictive success is a reliable indicator of approximate truth is consistent with there being some predictively successful but dead theories (Mizrahi 2013a, p. 3224).[8] In that respect, to argue that the phlogiston theory (or any other predictively successful but dead scientific theory, for that matter) is a counterexample to scientific realism is like arguing that single-stranded DNA is a counterexample to the double-helical model of DNA.[9] Both arguments fundamentally misunderstand the targets they aim to take down. The double-helical model of DNA is not supposed to be a universal statement about DNA and the realist thesis that predictive success is a reliable indicator of approximate truth is not supposed to be a universal generalization about scientific theories.[10] For these reasons, if the "Graveyard" Argument is meant to be an argument from counterexamples against scientific realism, then it cannot be said to be a sound argument.

## 5.2 The Positive Argument for Constructive Empiricism

As we have seen in Chap. 3, Constructive Empiricism (see Sect. 3.3) is supposed to be an antirealist alternative to realist positions about science. Unlike scientific realists, who tend to think that we have good reasons to believe that our best (that is, explanatorily and predictively successful) scientific theories are approximately true, constructive empiricists think that we should *accept* our best scientific theories, that

---

[8] Stathis Psillos (2018) puts this point as follows: "The relation between success and (approximate) truth, in this sense, is more like the relation between flying and being a bird: flying characterizes birds even if kiwis do not fly. If this is so, then there is need for more than one counter-example for the realist thesis to be undermined."

[9] The DNA example is borrowed from Kitcher (1993, p 118).

[10] After all, scientific realists are always careful to talk about "our best scientific theories" rather than scientific theories in general. I will say more about this in Chap. 6, Sect. 6.4.

is, we should believe what those theories say about observables, but we should not *believe* what those theories say about unobservable entities, processes, and events. For constructive empiricists, to accept a scientific theory is to believe that it is empirically adequate, that is, to believe that what the theory says about observables is true, but to suspend judgment with respect to what the theory says about unobservables.

Like scientific realism, Constructive Empiricism has a Positive Argument of its own. (See Chap. 4, Sect. 4.1 for the Positive Argument for scientific realism.) This Positive Argument can be found in Bas van Fraassen's seminal (1980, p. 73) book.

van Fraassen, Bas C. 1980. *The scientific image.* New York: Oxford University Press.

> There is also a positive argument for constructive empiricism—it makes better sense of science, and of scientific activity, than realism does and does so without inflationary metaphysics.

As I understand it, the argument that van Fraassen is making in this passage can be stated in canonical (or standard) form as follows:

(P1) Scientific activity involves positing unobservable entities, processes, and events in order to explain and predict observable phenomena.

(P2) The best explanation for scientific activity is Constructive Empiricism.

(P3) No hypothesis (for example, Explanationist Realism) explains scientific activity as well as Constructive Empiricism does.

Therefore,

(C) Constructive Empiricism is true.

Like the Positive Argument for scientific realism (see Chap. 4, Sect. 4.1), the Positive Argument for Constructive Empiricism is an instance of what philosophers of science call "Inference to the Best Explanation" (IBE).[11] As James Ladyman (2002, p. 209) points out, IBE "is sometimes also known as 'abduction'—following the terminology of Charles Peirce." However, some philosophers have argued that IBE and Peirce's abduction are different forms of inference (Douven 2017).[12] Be that as it may, for Peirce, abduction is a non-deductive form of inference. Likewise, IBE is typically construed as an ampliative, or non-deductive, form of argumentation that proceeds from a phenomenon that requires an explanation to the conclusion that the best explanation for that phenomenon is probably true.[13] As Ladyman (2007, p. 341) describes it, "Inference to the best explanation (IBE) is a (putative) rule of inference according to which, where we have a range of competing

---

[11] The phrase was coined by Gilbert Harman (1965).

[12] Daniel Campos (2011, p. 419) argues against the "tendency in the philosophy of science literature to link abduction to the inference to the best explanation (IBE), and in particular, to claim that Peircean abduction is a conceptual predecessor to IBE."

[13] For example, Alan Baker (2010, pp. 37–38) defines IBE as "A method of reasoning, also known as *abduction*, in which the truth of an hypothesis is inferred on the grounds that it provides the best explanation of the relevant evidence. In general, inference to the best explanation (IBE) is an ampliative (i.e., non-deductive) method" (emphasis in original).

hypotheses all of which are empirically adequate to the phenomena in some domain, we should infer the truth of the hypothesis which gives us the best explanation of those phenomena." The general form of IBE can be stated as follows:

1. Phenomenon $P$.
2. The best explanation for $P$ is $E$.
3. No other explanation explains $P$ as well as $E$ does.
4. Therefore, (probably) $E$.[14]

In the case of the Positive Argument for Constructive Empiricism, the phenomenon that demands an explanation is the fact that practicing scientists posit the existence of unobservable entities, processes, and events (for example, genes, genetic mutation, and speciation) in order to explain and predict observable phenomena (for example, biodiversity, fossils, and extinction). This scientific activity is supposed to be explained by the hypothesis that practicing "Scien[tists] aim [...] to give us theories that are empirically adequate" (van Fraassen 1980, p. 12). If Constructive Empiricism does indeed provide the best explanation for the fact that practicing scientists posit the existence of unobservables in order to explain and predict observable phenomena, then that would be a good reason to believe that Constructive Empiricism is probably correct. In other words, the premises of the Positive Argument for Constructive Empiricism, if true, would provide strong inductive support for its conclusion. Given that the Positive Argument for Constructive Empiricism is not meant to be a deductive argument, the next question is not whether it is valid or invalid, but rather whether it is strong or weak.

Scientific realists might protest against the Positive Argument for Constructive Empiricism by claiming that IBE is not a form of argumentation that is available to constructive empiricists. This is because constructive empiricists are generally critical of the use of IBE in science. As we have seen in Chap. 4 (Sect. 4.1), Bas van Fraassen (1980) takes issue with IBE. According to van Fraassen (1980, p. 143):

> [IBE] is a rule that selects the best among the historically given hypotheses. We can watch no contest of the theories we have so painfully struggled to formulate, with those no one has proposed. So *our selection may well be the best of a bad lot* (emphasis added).

According to constructive empiricists, then, from the fact that an explanation is the best one we could come up with, it does not follow that the explanation is probably true. For we might be working with a bad lot of explanations and the likely true explanation simply did not occur to us. If van Fraassen is right about this, then any inference to the best explanation would be a weak argument. For, even if no other explanation explains $P$ as well as $E$ does, $E$ would still not be more probable or likely to be true if $E$ is simply the best of a "bad lot" of explanations.[15] If constructive empiricists are right about this, then one could apply the same reasoning to the Positive Argument for Constructive Empiricism. That is to say, when constructive

---

[14] For more on the structure of IBE, see Psillos (2007).

[15] We will revisit van Fraassen's criticism against IBE as inference from a "bad lot" of explanations in Chap. 6, Sect. 6.4.

empiricists infer that Constructive Empiricism is better than realist positions about science on explanatory grounds, since Constructive Empiricism supposedly "makes better sense of science, and of scientific activity, than realism does and does so without inflationary metaphysics" (van Fraassen 1980, p. 73), they may simply be selecting the best hypothesis among several bad ones. If Constructive Empiricism is one hypothesis among several bad ones, then we have no good reasons to believe it, by the constructive empiricist's own lights. For this reason, even constructive empiricists cannot say that their own Positive Argument for Constructive Empiricism is a strong inductive argument, given their own objection to the use of IBE in science.[16]

For the sake of argument, however, let us grant that IBE is a legitimate form of non-deductive (or inductive) argumentation. The next question, then, is whether the premises of the Positive Argument for Constructive Empiricism are in fact true? Is the Positive Argument for Constructive Empiricism a cogent argument? Premise (P1) is not in question, of course, for it states the phenomenon to be explained. So, to determine the cogency of the Positive Argument for Constructive Empiricism we need to ask the following question: Does Constructive Empiricism provide the best explanation for the phenomenon of scientific activity (that is, scientists positing unobservable entities, processes, and events in order to explain and predict observable phenomena)? Or are there alternative explanations that explain this phenomenon just as well as Constructive Empiricism does?

Scientific realists would object to (P3) of the Positive Argument for Constructive Empiricism. They would argue that realist positions, such as Explanationist Realism, can explain scientific activity just as well as (if not better than) Constructive Empiricism can. On Explanationist Realism (see Chap. 3, Sect. 3.1), for example, practicing scientists posit unobservables to explain and predict observable phenomena not merely because they want to "save the phenomena," or get the observable facts right, but also because they want approximately true theories, or get the unobservable facts right, too. When scientific theories make predictions that are borne out by repeated experimentation and observation, then we can be quite confident that those theories are approximately true. For, if our best scientific theories were not even approximately true, they would not have been successful in the first place (see Chap. 4, Sect. 4.1). If this is correct, then (P3) of the Positive Argument for Constructive Empiricism cannot be said to be true as scientific realists would argue that there are alternative explanations for the phenomenon of scientific activity that can explain it just as well as (if not better than) Constructive Empiricism can. And if there are alternative explanations for the phenomenon of scientific activity that can explain it just as well as Constructive Empiricism can, then the Positive Argument for Constructive Empiricism, even if strong, cannot be said to be a cogent argument.

---

[16]For more on this problem with the Positive Argument for Constructive Empiricism, see Mizrahi (2018).

## 5.3  The Underconsideration Argument

As we have seen in Sect. 5.1, some antirealists are pessimistic about the prospects of our current scientific theories. That is to say, they think that our current scientific theories will likely end up in the so-called "historical graveyard of science" (Frost-Arnold 2011, p. 1138). Other antirealists are also pessimistic about the epistemic credentials of scientists (as opposed to the epistemic prospects of scientific theories). That is to say, they argue that scientists are not in an epistemic position to discover scientific truths in the way that scientific realists claim that they are. This antirealist claim, namely, that scientists do not occupy an epistemically privileged position such that they can tell which of the theories they are testing are (approximately) true and which are not, is then used as a premise in an antirealist argument known as the "Underconsideration Argument." A persistent defender of this antirealist argument is Brad Wray (2008, p. 317).

Wray, Brad, K. 2008. The argument from underconsideration as grounds for antirealism: a defence. *International Studies in the Philosophy of Science* 22 (3): 317–326.

> *when scientists evaluate theories they only ever consider a subset of the theories that can account for the available data, specifically, those theories that have been developed. As a result, the anti-realist argues, when a scientist judges one theory to be superior to competitor theories she is hardly warranted in drawing the conclusion that the superior theory is likely true with respect to what it says about unobservable entities and processes. Anti-realists claim that the inference to the likely truth of the superior theory presumes scientists are especially skilled at developing theories that are apt to be true. But the history of science seems to suggest otherwise: scientists are not so privileged.*

As I understand it, the argument that Wray is making in this passage can be stated in canonical (or standard) form as follows:

(P1) Scientists are warranted in concluding that the theory that is judged to be superior to a few other theories from a subset of competing theories is likely true with respect to what it says about unobservables only if scientists are epistemically privileged (that is, scientists are skilled at developing theories that are apt to be likely true).

(P2) It is not the case that scientists are epistemically privileged.

Therefore,

(C) It is not the case that scientists are warranted in concluding that the theory that is judged to be superior to a few other theories from a subset of competing theories is likely true with respect to what it says about unobservables.[17]

Reconstructed in this way, the Underconsideration Argument is a deductive argument (in particular, a *modus tollens*: If *A*, then *B*, not *B*; therefore, not *A*). Since the premises of Wray's Underconsideration Argument, namely, (P1) and (P2), successfully provide logically conclusive support for the conclusion, namely, (C), this

---

[17] Cf. Brad Wray's (2012, p. 377) reconstruction of the Underconsideration Argument. See also Wray (2018, p. 43).

argument can be said to be valid. The next question, then, is whether the premises are in fact true. Is the Underconsideration Argument sound?

The key premise of the Underconsideration Argument seems to be (P2). Before we can say whether (P2) is true or false, we need to understand what it means. What does it mean to say that scientists are (or are not) epistemically privileged? To be privileged is to have special rights or advantages. In that case, talk of scientists not being "epistemically privileged" could mean that scientists do not have or do not enjoy special epistemic rights or advantages. What sort of special epistemic rights or advantages can scientists claim to have or enjoy in the first place? Well, they can claim to have special training and, as a result, certain skills that allow them to investigate specific domains in nature. In that respect, trained scientists have special *epistemic advantages* over untrained laymen in terms of having the knowledge and skills to investigate nature by means of observation and experimentation. Scientists can also claim to enjoy special access to advanced technology, such as observation instruments and experimentation techniques, and to having the advantage of using these technologies in their investigations of nature. In that respect, trained scientists have special *epistemic rights* that untrained laymen lack in terms of being entitled to use advanced instruments of observation and experimentation, such as particle accelerators, electron microscopes, and DNA sequencers, in their investigations of the natural world. On this reading of "epistemically privileged," then, why is it that scientists are not epistemically privileged? Given their unique training, knowledge, skills, and access to advanced technology, why should we think that scientists are not "skilled at developing theories that are apt to be true"? Presumably, because they never get better at what they do. That is to say, all the scientific training, knowledge, skills, and access to advanced technology in the world could never make scientists sufficiently skilled at developing theories that are approximately true. But why assume that? That is to say, (P2) of the Underconsideration Argument seems to presuppose that the aforementioned aspects of theory generation, such as scientific training, technical skills, advanced instrumentation, and the like, do not change over time and that scientists never get better at what they do. But this is an assumption that needs to be argued for rather than taken for granted.

In response, antirealists could appeal to the history of science, and argue, as Wray (2018, p. 43) in fact does, that "the history of science seems to suggest [that] [s]cientists [do] not have [...] epistemic privilege." Indeed, Wray (2012, p. 380) writes that the "no-privilege thesis [that is, (P2) of the Underconsideration Argument] asks us to acknowledge the similarities between current scientists and their predecessors." He quotes Mary Hesse (1976, p. 266), who argues that the support for the no-privilege premise, that is, (P2) of the Underconsideration Argument, comes from an "induction from the history of science." In that case, antirealists would be making an inductive argument, similar in structure to the "Graveyard" Argument (see Sect. 5.1), in support of (P2) of the Underconsideration Argument. This inductive argument can be stated in canonical (or standard) form as follows:

(P) Many past scientists were not epistemically privileged (that is, they were not skilled at developing theories that are apt to be likely true).

Therefore,

(C) Current (and future) scientists are not epistemically privileged.

If this inductive argument were cogent, it would provide probable support for (P2) of the Underconsideration Argument. For, if current (and future) scientists are just like past scientists, then they, too, are not epistemically privileged, just as past scientists were not epistemically privileged. But are current (and future) scientists really like past scientists? Or are there relevant differences between the two such that no strong inductive inferences can be drawn from what was true about past scientists to what is true about current scientists (and what will be true about future scientists)? That is to say, if current (and future) scientists are not quite like past scientists, then, even if past scientists were not epistemically privileged, the conclusion that current (and future) scientists are not epistemically privileged as well would not follow inductively from that because the sample of scientists in this inductive generalization from a sample would not be sufficiently uniform for projecting the property of "being epistemically unprivileged" from past scientists to current (and future) scientists. In other words, if current (and future) scientists differ from past scientists in relevant respects, then they may also differ in terms of being epistemically (un)privileged.

Now, let us consider some of the relevant respects in which current scientists are unlike past scientists (Mizrahi 2013b, p. 398):

- Current scientists learn from their predecessors' successes and failures and seek to avoid their predecessors' mistakes. For example, scientists have learned from the mistakes of Franz Joseph Gall and others, which had led to the development of phrenology. Nowadays, scientists would not make the mistake of supposing that there is a simple relationship between the relative size of different parts of a person's brain and that person's personality and character traits.
- Current scientists have access to methods and technologies that were not available to their predecessors. For example, Galileo and his contemporaries may have had telescopes, but they did not have radio telescopes, infrared telescopes, microwave anisotropy probes, and much more.
- Current scientists are able to collaborate with colleagues on increasingly large-scale, costly, and international projects. For example, it is difficult to see how large-scale, international scientific projects, such as CERN's Large Hadron Collider, the Human Genome Project, the Event Horizon Telescope, and the Human Brain Project, could have been executed even just a few decades ago.

Of course, current scientists are also like past scientists in some respects. For example, both current and past scientists present their research to their academic peers, and both publish their results in academic books and professional journals. But these respects in which they are alike are not relevant to being epistemically privileged, that is, to developing theories that are likely to be approximately true, and hence to the inductive inference from the failures of past scientists to the claim that current (and future) scientists are likely to fail, too. This is so precisely because current (and future) scientists have an expanding track record of successes and failures to draw upon and learn from. In other words, just as "the manifest difference in methods between us and our predecessors breaks the pessimist's induction," according to Sherri Roush (2010, p. 55; see Sect. 5.1), the aforementioned differences between current scientists and their predecessors break the inductive inference in support of (P2) of the Underconsideration Argument. Moreover, even if we grant that past scientists were not epistemically privileged (that is, they were not skilled at developing theories that are apt to be likely true), as Wray (2008, p. 317) argues, "nothing follows from this about whether we have a right to our confidence in our particular theories" (Roush 2010, p. 55) unless the antirealist can also show that the lack of epistemic privilege of past scientists is a reason to think that current scientists lack epistemic privilege as well, which has not been shown, since the manifest differences in scientific knowledge, methods, instruments, and technologies between current scientists and their predecessors break the inductive inference in support of (P2) of the Underconsideration Argument.

As we have seen in Sect. 5.1, for scientific realists, the problem with the "Graveyard" Argument is that it overemphasizes the similarities and underemphasizes the dissimilarities between current scientific theories and their predecessors. This is why, scientific realists argue, the failure of past *theories* does not warrant a strong inductive inference to the failure of current (and future) *theories*. Similarly, as far as scientific realists are concerned, the problem with the inductive argument in support for (P2) of the Underconsideration Argument is that it overemphasizes the similarities and underemphasizes the dissimilarities between current scientists and their predecessors. Current scientists learn from their predecessors and they seek to avoid their predecessors' mistakes. Furthermore, current scientists have access to methods and technologies that were not available to their predecessors. For scientific realists, these aspects of scientific change make a difference insofar as the ability of scientists to develop theories is concerned. This is why, scientific realists would argue, the failure of *past scientists* does not warrant a strong inductive inference to the failure of current (and future) *scientists*. As Alexander Bird (2007a, p. 80) puts it:

> Later scientific *theories* are not invented independently of the successes and failures of their predecessors. New *theories* avoid the pitfalls of their falsified predecessors and seek to incorporate their successes (emphasis added).

Similarly, later *scientists* are not trained independently of the successes and failures of their predecessors. Current (and future) scientists avoid the pitfalls of their predecessors and seek to incorporate their successes. If this is correct, then the inductive argument that purports to provide support for (P2) of the Underconsideration Argument is a weak argument. Without strong inductive support for (P2), then, this premise of the Underconsideration Argument cannot be said to be true, and thus the Underconsideration Argument itself, although valid, cannot be said to be sound.

# 5.4 The Argument from Unconceived Alternatives

Much like the Underconsideration Argument (see Sect. 5.3), the Argument from Unconceived Alternatives, is an argument about *scientists* rather than scientific *theories*. The Underconsideration Argument proceeds from the premise that scientists are not epistemically privileged, whereas the Argument from Unconceived Alternatives proceeds from the premise that past scientists were not in a position to conceive of all the alternatives to the few theories they actually ended up testing. Subsequently, it turned out that those alternative theories, which were unconceived at the time, were equally well-confirmed by the evidence available at the time. From this, then, it is supposed to follow that current (and future) scientists are not in a position to conceive of all the alternatives to the few theories they actually end up testing, either. Therefore, the argument goes, it will probably turn out that those alternative theories, which are unconceived at the time, will be equally well-confirmed by the evidence available at the time, and thus that we have no good reasons to believe that our current best theories are approximately true.

This argument was made by Kyle Stanford on several occasions. According to Stanford (2006, p. 19), "the history of scientific inquiry itself offers a straightforward rationale for thinking that there typically are alternatives to our best theories equally well-confirmed by the evidence, even when we are unable to conceive of them at the time." Stanford calls this the "Problem of Unconceived Alternatives" (PUA). For him, the PUA provides the inductive basis for "the following New Induction over the History of Science" (Stanford 2001, p. S9).

Stanford, Kyle P. 2001. Refusing the devil's bargain: what kind of underdetermination should we take seriously? *Philosophy of Science* 68 (S3): S1–S12.

> *that we have, throughout the history of scientific inquiry and in virtually every field, repeatedly occupied an epistemic position in which we could conceive of only one or a few theories that were well-confirmed by the available evidence, while subsequent history of inquiry has routinely (if not invariably) revealed further, radically distinct alternatives as well-confirmed by the previously available evidence as those we were inclined to accept on the strength of that evidence.*

As I understand it, the argument that Stanford is making in this passage can be stated in canonical (or standard) form as follows:

(P) Past scientists typically failed to conceive of alternatives to the few theories that they confirmed by the evidence available to them at the time.

Therefore,

(C) Current (and future) scientists fail to conceive of alternatives to the few theories that they confirm by the evidence available to them at the time.

If (P) provides strong inductive support for (C), then we should not believe that our current best theories are approximately true because there probably are many alternative theories, which are equally well-confirmed by the available evidence, but that we have not even conceived of yet (Magnus 2010, p. 807). In that respect, this New Induction over the History of Science, or Argument from Unconceived

Alternatives, is supposed to be an inductive argument. So the next question is whether this argument is strong or weak.[18]

As we have seen in Sects. 5.1 and 5.3, antirealist arguments that seek to draw conclusions from the history of science by induction tend to assume that past scientists and/or scientific theories are like current (and future) scientists and/or scientific theories, so what was true of past scientists and/or scientific theories is probably true of present scientists and/or scientific theories, and will probably be true of future scientists and/or scientific theories as well. But is this a warranted assumption. After all, scientists are human beings and, as any good social scientist knows, human beings are not only "much more complicated than anything else studied by the methods of natural science" (Rosenberg 1980, p. 93) but also differ greatly from one human society or group to another. Scientists are human beings, of course, and so they, too, are "much more complicated than moving bodies, chemical reagents, and ocean tides" (Rosenberg 1980, p. 10). For these reasons, no simple inductive inferences can be made from what is true about some scientists to what is true (or will be true) about *other* scientists. In other words, when we are dealing with human subjects, we cannot simply make what Peter Godfrey-Smith (2011, p. 41) calls "seen one, seen them all" inductive inferences. According to Godfrey-Smith (2011, p. 41), a "seen one, seen them all" inductive inference is when "one instance of an *F* would be enough, in principle, if you picked the right case and analyzed it well." This is because, as far as "seen one, seen them all" inductive inferences are concerned, "sample size per se does not matter, randomness does not matter, but the status of the kinds matters enormously" (Godfrey-Smith 2011, p. 42). In other words, if we have reasons to believe that our instance of an *F* belongs to a certain *kind* of things—what philosophers of science sometimes call a "natural kind"— such that all *F*s are the same *kind* of thing, then the inductive inference from "this is an *F* that is also a *G*" to "All *F*s are *G*s" would be a strong inductive inference. For example, all copper rods have the same atomic structure, which is why if we have seen one copper rod, we have seen them all. Accordingly, from testing just a few copper rods for electrical conductivity, we can inductively infer that all copper rods conduct electricity. That is to say, the inductive inference from "these few copper rods conduct electricity" to "all copper rods conduct electricity" is a strong inductive inference because of the uniform atomic structure of copper rods. Copper rods are a natural kind, if you will. Scientists, however, are not like copper rods. Scientists do not have a uniform atomic structure. Scientists are not a natural kind. Consequently, when it comes to scientists, no strong "seen one, seen them all" inductive inferences can be made based on what holds for just a few scientists. For example, if we study a few past scientists and learn that they had failed to conceive of alternatives to the few theories that they confirmed by the evidence available to them at the time, no strong inductive conclusions about other scientists would

---

[18] Darrell Rowbottom (2019) extends the Argument from Unconceived Alternatives to include aspects of science in addition to scientific theories, such as models, predictions, instruments, and more.

follow from this small sample. (Recall our discussion of the use of case studies as evidence from Chap. 2, Sect. 2.2.)

When we are dealing with complicated things, such as scientists, which do not belong to a natural kind, we need a different inductive bridge from evidence or premises to conclusions. This inductive bridge from premises to conclusion is generalization from random samples. As Godfrey-Smith (2011, p. 42) puts it, "This form of inference has the following features: sample size matters, randomness matters, and 'law-likeness' or 'naturalness' does not matter." That is to say, when we are dealing with diverse individuals, we cannot generalize from one or two individuals by assuming that what is true for one or two is true for the others as well. This is because the "seen one, seen them all" inference does not work for things that do not belong to a uniform, natural kind. When we are dealing with diverse individuals, and we want to reason inductively about them, we need to make sure that we reason inductively from large and random samples, if we want our inductive inferences to be strong inferences. As far as the Argument from Unconceived Alternatives is concerned, however, the sample of scientists is small. Stanford (2006, pp. 19–20) provides only eight examples of allegedly unconceived alternatives.

Moreover, the sample of scientists was not randomly selected, but rather was cherry-picked in order to advance an argument against scientific realism. The lack of random sampling undermines the inductive inference from a sample of past scientists to what is true about current scientists (and what will be true about future scientists) because it biases the sample on which the inductive inference is based. When a sample is biased, the door is then open to all kinds of mistakes in inductive reasoning, such as confirmation bias and Type I errors. Confirmation bias occurs when one selects and gives more weight to any evidence that supports one's previously held beliefs. Clearly, any inductive inference made from evidence carefully selected to support one's previously held beliefs will likely be non-cogent. A Type I error occurs when one accepts a positive result when that positive result is in fact false. This is why Type I errors are also referred to as "false positives." For example, like any other tests for viral infections, viral tests for Coronavirus Disease 2019 (COVID-19) can have false positive and false negative results. Suppose that a healthy person, who is not infected with the Severe Acute Respiratory Syndrome Coronavirus 2 (SARS-CoV-2), gets tested and the viral test returns a positive result, that is, a result that the person is infected with SARS-CoV-2. This is a false positive because the test result is positive even though the person is not infected. When we accept the false positive result of this test, we make a Type I error. On the other hand, suppose that a sick person, who is infected with SARS-CoV-2, gets tested and the viral test returns a negative result, that is, a result that the person is not infected with SARS-CoV-2. This is a false negative because the test result is negative even though the person is infected. When we accept the false negative result of this test, we make a Type II error.

Now, without random sampling, we cannot tell if Stanford's sample of past scientists is not simply a biased sample. Perhaps, even unbeknownst to him, Stanford has selected and gave more weight to any evidence that supported his previously held beliefs. Furthermore, without random sampling, we cannot tell if Stanford's sample of past scientists is not simply a sample of false positives, that is, a sample of past scientists who are thought to have failed to conceive of alternative theories but in fact did not fail to do so.[19] Perhaps, even unbeknownst to him, Stanford has made a Type I error. That is to say, he examined the history of science and "tested" past scientists for whether they succeeded or failed to conceive of alternative theories, and his tests returned false positive results, that is, past scientists who are thought to have failed to conceive of alternative theories but in fact did not fail to do so. Since Stanford's sample of past scientists was not randomly selected, but rather was cherry-picked in order to advance an argument against scientific realism, it is likely that such errors were made. If the sample from which the conclusion of the Argument from Unconceived Alternatives is drawn is too small and not random, then no strong inductive generalizations can be made based on that sample. In other words, the Argument from Unconceived Alternatives is a weak inductive argument.

Indeed, as we have seen in Sect. 5.3, there are reasons to think that current scientists are in fact unlike past scientists in many relevant respects. Current scientists learn from their predecessors and they seek to avoid their predecessors' mistakes. Furthermore, current scientists have access to knowledge, methods, instruments, and technologies that were not available to their predecessors. In other words, just as "the manifest difference in methods between us and our predecessors breaks the pessimist's induction," according to Sherri Roush (2010, p. 55; see Sect. 5.1), the manifest differences in scientific knowledge, methods, instruments, and technologies between current scientists and their predecessors break the inductive Argument from Unconceived Alternatives. Moreover, even if we grant that past scientists typically failed to conceive of alternatives to the few theories that they confirmed by the evidence available to them at the time, as Stanford (2001, p. S9) argues, "nothing follows from this about whether we have a right to our confidence in our particular theories" (Roush 2010, p. 55) unless antirealists can also show that the lack of theoretical imagination of past scientists is a reason to think that current scientists lack theoretical imagination as well, which has not been shown, since the manifest differences in scientific knowledge, methods, instruments, and technologies between current scientists and their predecessors break the inductive Argument from Unconceived Alternatives. If this is correct, then one might complain that the Argument from Unconceived Alternatives overemphasizes the similarities and underemphasizes the dissimilarities between current scientists and their predecessors. Given that there are relevant differences between past scientists and current scientists (see Sect. 5.3), the inductive inference from what was true of past scientists to what is true of present scientists, and what will be true of future scientists, is weak. In other words, the Argument from Unconceived Alternatives is a weak inductive argument.[20]

---

[19]In Mizrahi (2015), I argue that some of the examples in Stanford's sample are in fact false positives.

[20]In Mizrahi (2015), I argue that the historical evidence Stanford cites in support of his New Induction is indeterminate between a pessimistic (antirealist) interpretation and an optimistic

There is another problem with the Argument from Unconceived Alternatives that makes it a problematic argument for antirealists to use as an argument against scientific realism. Recall that, according to Stanford (2001, p. S9), the "Problem of Unconceived Alternatives" (PUA) applies to "virtually every field." That is to say, in every field of inquiry, inquirers often occupy an epistemic position in which they could conceive of only a few theories that are well-confirmed by the available data but could not conceive of alternative theories that subsequent history of inquiry would show to be equally well-confirmed by the available evidence. Since this is supposedly true of *every field of inquiry*, it is supposed to be true of philosophy as well, given that philosophy is a field of inquiry. In that case, philosophers, too, often occupy an epistemic position in which they could conceive of only a few theories that are well-confirmed by the available evidence but could not conceive of alternative theories that subsequent history of philosophical inquiry would show to be equally well-confirmed by the available evidence. This, in turn, means that Stanford's New Induction applies not only to science but also to philosophy. Accordingly, one could make a Stanford-like New Induction over the History of Philosophy. Such an argument can be stated in canonical (or standard) form as follows:

(P) Past philosophers typically failed to conceive of alternatives to the few theories that they confirmed by the evidence available to them at the time.

Therefore,

(C) Current (and future) philosophers fail to conceive of alternatives to the few theories that they confirm by the evidence available to them at the time.

If this inductive argument were cogent, however, then it would mean that scientific antirealism is probably one of the few theories that philosophers could conceive of and that were well-confirmed by the available evidence, but that subsequent history of philosophical inquiry would show that there are alternatives to antirealism that philosophers could not conceive of and that are equally well-confirmed by the available evidence. In other words, if we should not believe that our current best scientific theories are approximately true because there probably are many alternative scientific theories, which are equally well-confirmed by the available evidence, but that scientists have not even conceived of yet (Magnus 2010, p. 807), then we should not believe that our current philosophical theories are approximately true because there probably are many alternative philosophical theories, which are equally well-confirmed by the available evidence, but that philosophers have not even conceived of yet (Mizrahi 2016b, pp. 60–63).

Accordingly, if antirealists would like to use Stanford's New Induction as an argument against scientific realism, then they would have to concede that a Stanford-like New Induction applies to philosophical inquiry as well, given that the PUA applies to "virtually every field" (Stanford 2001, p. S9), including philosophy. Again, as Stanford (2006, p. 44) himself puts it:

---

(realist) interpretation. If the historical evidence is indeterminate between scientific realism and antirealism, then it cannot be used to argue in favor of one over the other (see Chap. 4, Sect. 4.5).

the problem of unconceived alternatives and the new induction suggest not that present theories are no more likely to be true than past theories have turned out to be, but instead that *present theorists are no better able to exhaust the space of serious, well-confirmed possible theoretical explanations of the phenomena than past theories have turned out to be* (emphasis added).

Of course, insofar as philosophers are theorists as well, the PUA applies to them because it applies to every theorist. That is to say, present philosophers, too, are *no better able to exhaust the space of serious, well-confirmed possible theoretical explanations of the phenomena than past philosophers have turned out to be.* Granting that a Stanford-like New Induction applies to philosophical inquiry, however, means that we should not believe that our current philosophical theories are (approximately) true because there probably are many alternative philosophical theories, which are equally well-confirmed by the available evidence, but that philosophers have not even conceived of yet (Mizrahi 2016b, pp. 60–63).[21] From this, in turn, it follows that we should not believe scientific antirealism, since antirealism is probably one of those current philosophical theories with unconceived alternatives that are equally well-confirmed by the available evidence. If this is correct, then, by advancing Stanford's New Induction as a cogent argument against realism about theories, antirealists would be undermining their own theory, namely, antirealism. For this reason, it seems that even antirealists cannot endorse the Argument from Unconceived Alternatives as a cogent argument against scientific realism, for, in doing so, they would be undermining their own antirealist position. So, even if it were a strong argument, the Argument from Unconceived Alternatives cannot be said to be a cogent argument against scientific realism.

## 5.5   The Argument from Changing Research Interests

As we have seen in Sect. 5.1, antirealists tend to subscribe to the view that the history of science is a "graveyard of dead epistemic objects" (Chang 2011, p. 426), that is, discarded theories and abandoned theoretical posits.[22] From this historical understanding of science as a "graveyard of dead epistemic objects" (Chang 2011, p. 426), antirealists then draw the conclusion that current (and future) scientific theories and theoretical posits will end up in that "graveyard" as well (see Sect. 5.1 on the "Graveyard" Argument). As many antirealists do, Brad Wray (2018) also thinks that our current scientific theories will be discarded in the future, but he develops an original argument for this conclusion that, for him, explains why the history of

---

[21] Cf. Sterpetti (2019). I reply to Sterpetti (2019) in Mizrahi (2019).

[22] As Peter Lipton (2005, p. 1265) puts it, "The history of science is a graveyard of theories that were empirically successful for a time, but are now known to be false, and of theoretical entities—the crystalline spheres, phlogiston, caloric, the ether and their ilk—that we now know do not exist." In Mizrahi (2016a), I provide empirical evidence against this "graveyard" picture of the history of science. Cf. Tulodziecki (2017).

science is marked by discarded or failed theories. According to Wray (2018, p. 187), "the research interests that determine what sorts of issues a scientist investigates" change over time. For example, "At one point in the history of astronomy, astronomers were concerned with the question of whether or not planets were self-illuminating" (Wray 2018, p. 187). For current astronomers, Wray says, this question is no longer a concern. He takes this to be evidence against scientific realism or a reason "to expect that many of the theories we currently accept, despite their many impressive successes, will be discarded sometime in the future" (Wray 2019, p. 555). For Wray, this explains why the history of science is marked by failed theories, the so-called "historical graveyard of science" (Frost-Arnold 2011, p. 1138). He makes this argument in his book-length defense of antirealism (Wray 2018, p. 187).

Wray, Brad K. 2018. *Resisting scientific realism.* Cambridge: Cambridge University Press.

> *Every theory is only ever a partial representation of the world, thus every theory leads scientists to disregard some features of the world. Scientists' interests determine which features they disregard in their theories, and as they realize their research goals, their interests will change. Consequently, a theory that effectively served the interests of scientists at one time is apt to seem inadequate at some later time, when scientists have different research interests. At this later time, the theory is vulnerable to being discarded and replaced by a new theory that better serves current research interests.*

As I understand it, the argument that Wray is making in this passage can be stated in canonical (or standard) form as follows:

(P1) If scientific theories are partial representations of the world, then scientists have to decide which features of the world to disregard in their theories and their decisions are determined by their research goals and interests.
(P2) Scientific theories are partial representations of the world.

Therefore,

(C1) Scientists have to decide which features of the world to disregard in their theories and their decisions are determined by their research goals and interests. [from (P1) & (P2)]
(P3) If scientists have to decide which features of the world to disregard in their theories, and their decisions are determined by their research goals and interests, then their theories will be replaced by new theories as they realize their research goals and their research interests change.

Therefore,

(C2) Scientific theories will be replaced by new theories as scientists realize their research goals and their research interests change. [from (C1) & (P3)]

This line of reasoning is valid. That is to say, at each step, the premises provide logically conclusive support for the conclusion that follows from those premises. The next question, then, is whether the premises are actually true. Is Wray's Argument from Changing Research Interests sound?

A key premise of Wray's Argument from Changing Research Interests is (P3). It assumes that the research goals and interests of scientists change over time. This assumption is then used as a premise in support of the conclusion that scientific theories will change, that is, replaced by new theories, with the changing research goals and interests of scientists. Now, it may be the case that secondary, small-scale, or short-term research goals do change over time, as Wray claims, but perhaps there are primary, large-scale, or long-term research goals that do not change over time. For example, perhaps ancient astronomers aimed at answering "the question of whether or not planets were self-illuminating," as Wray (2018, p. 187) says, whereas current astronomers have no such research goal or interest. But these may be secondary, small-scale, or short-term research goals and interests, for example, "Are planets self-illuminating?" or "Is there liquid water on the surface of Mars?" Perhaps ancient astronomers and current astronomers share primary, large-scale, or long-term research goals and interests, such as "What is the most accurate model of our solar system?" or "What is the best definition of 'planet'?" Current astronomers aim to answer such questions, which date back to the origins of astronomy, as the following quote from David Weintraub's book, *Is Pluto a Planet?* (2007, p. 220) illustrates:

> Our quest to answer the question *Is Pluto a planet?* led us directly to a question about physics: *What is a planet?* Answering this second question, which was not simple or easy, has revealed that we live in a solar system that is quite different from the one we thought we lived in: *The solar system has more than twenty planets!* (emphasis in original)

Accordingly, if there are such primary, large-scale, or long-term research goals and interests that remain relatively fixed over time, then it does not necessarily follow that scientific theories will change with the changing research goals and interests of scientists from the fact that there are *also* secondary, small-scale, or short-term research goals and interests that do change over time. For the claim about changing research goals and interests fails to distinguish between the primary and the secondary research goals and interests of scientists.

In other words, (P3) of Wray's Argument from Changing Research Interests is ambiguous. More specifically, the phrase "research goals and interests" is ambiguous between two interpretations: primary, large-scale, or long-term research goals and interests versus secondary, small-scale, or short-term research goals and interests. On the one hand, if the phrase "research goals and interests" means secondary, small-scale, or short-term research goals and interests, then change in such research goals and interests does not entail that scientific theories will be replaced by new theories as scientists realize their secondary research goals and their secondary research interests change because scientists might still have primary, large-scale, or long-term research goals and interests that remain relatively fixed over time. On the other hand, if the phrase "research goals and interests" means primary, large-scale, or long-term research goals and interests, then Wray does not provide sufficient evidence or reasons to believe that such primary research goals and interests do change over time.

In fact, as we have seen in Chap. 3, there are both realist and antirealist positions in the scientific realism/antirealism debate according to which scientists do have primary research goals and interests that are not supposed to change over time.

According to Constructive Empiricism (see Sect. 3.3), "Science aims to give us theories which are empirically adequate" (van Fraassen 1980, p. 12). For constructive empiricists, then, the primary research goal or aim of scientists is *empirical adequacy*. Empirical adequacy is a research goal or aim that is not supposed to change over time. If constructive empiricists are right about this, then (P3) of Wray's Argument from Changing Research Interests is false. Even if scientists have to decide which features of the world to disregard in their theories, and their decisions are determined by their secondary research goals and interests, it does not necessarily follow that their scientific theories will be replaced by new theories as their secondary research goals and interests change, for empirical adequacy will remain their primary research goal or aim, even as they realize their secondary research goals.

As we have also seen in Chap. 3, constructive empiricists contrast their antirealist position with scientific realism by saying that, according to scientific realism, "Science aims to give us, in its theories, a literally true story of what the world is like" (van Fraassen 1980, p. 8), rather than empirically adequate theories. For scientific realists, then, the primary research goal or aim of scientists is *approximate truth*. Approximate truth is a research goal or aim that is not supposed to change over time. Again, if scientific realists are right about this, then (P3) of Wray's Argument from Changing Research Interests is false. Even if scientists have to decide which features of the world to disregard in their theories, and their decisions are determined by their secondary research goals and interests, it does not necessarily follow that their theories will be replaced by new theories as their secondary research goals and interests change, for approximate truth will remain their primary research goal or aim, even as they realize their secondary research goals. For these reasons, (P3) of Wray's Argument from Changing Research Interests may be false, which means that this argument, albeit valid, cannot be said to be sound.

## 5.6  Summary

In the contemporary scientific realism/antirealism debate, the antirealist argument that has attracted the most attention from scientific realists and antirealists alike is the so-called "pessimistic induction" or "pessimistic meta-induction." According to this argument, the history of science is a graveyard of "once successful but now dead" scientific theories, and so our current theories will end up in the historical graveyard of science as well. Scientific realists object to this argument by pointing out that the few examples selected to support the "graveyard" premise of the argument are not representative of the general population of scientific theories because they are too few, and not randomly selected, but rather cherry-picked in an attempt to refute scientific realism. Moreover, selected counterexamples cannot refute scientific realism because a predictively successful but dead theory does not count as an effective counterexample against the claim that predictive success is a reliable indicator of approximate truth (see Sect. 5.1). Constructive empiricists make a positive argument for their position by arguing that Constructive Empiricism is the best

explanation for scientific activity. Scientific realists object by pointing out that constructive empiricists cannot help themselves to making Inferences to the Best
Explanation (IBE) and that there are realist explanations that explain scientific
activity just as well as Constructive Empiricism does (see Sect. 5.2). Two antirealist
arguments that focus on scientists, rather than on scientific theories, are the
Underconsideration Argument and the Argument from Unconceived Alternatives.
According to the first, scientists are not epistemically privileged, which is why we
should not believe that their best theories are approximately true (see Sect. 5.3).
According to the second, scientists routinely fail to conceive of alternatives to their
well-confirmed theories, even though those alternatives later end up being equally
well-confirmed by the available evidence (see Sect. 5.4). Scientific realists object to
these arguments by pointing out that there are relevant differences between past
scientists and current scientists, so we cannot simply infer from what was true about
past scientists to what is true about current scientists and what will be true about
future scientists by induction. Finally, according to the Argument from Changing
Research Interests, our best scientific theories are likely to be replaced by new theories as the research goals of scientists are achieved and their research interests
change (see Sect. 5.5). Scientific realists could object to this argument by distin-
guishing between primary and secondary research goals. The latter may change, but
the former could be stable, which means that theories could be stable as well.
Indeed, according to both scientific realist and antirealist positions in the scientific
realism/antirealism debate, scientists supposedly have primary research goals that
do not change over time. For scientific realists, approximate truth is such a primary
research goal, whereas for constructive empiricists, it is empirical adequacy.

## Glossary

**Antirealism** An agnostic or skeptical attitude toward the theoretical posits (that is,
   unobservables) of scientific theories. Antirealism comes in different varieties,
   such as Constructive Empiricism (see Chap. 3, Sect. 3.3) and Instrumentalism
   (see Chap. 3, Sect. 3.2).
**Approximate truth** Closeness to the truth or truthlikeness. To say that a theory is
   approximately true is to say that it is close to the truth. According to some scientific realists, approximate truth is the aim of science. (See Chap. 2, Sect. 2.1.)
**Case study** A particular, detailed description of a scientific activity, a scientific
   practice, or an episode from the history of science. (See Chap. 2, Sect. 2.2.)
**Cherry-picking** A sample from which an inductive inference is made is said to be
   cherry-picked when it is not randomly selected. (See Chap. 5, Sect. 5.1.)
**Constructive Empiricism** The view that the aim of science is to construct empirically adequate theories. A theory is empirically adequate when what the theory
   says about what is observable (by us) is true. (See Chap. 3, Sect. 3.3.)
**Empirical success** A scientific theory is said to be empirically successful just in
   case it is both explanatorily successful (that is, it explains natural phenomena

that would otherwise be mysterious to us) and predictively successful (that is, it makes predictions that are borne out by observation and experimentation). (See Chap. 3, Sect. 3.1.)

**Explanationist Realism** The view that realist commitments are warranted with respect to the theoretical posits that are responsible for—or can best explain—the predictive success of our best scientific theories (also known as "Deployment Realism"). (See Chap. 3, Sect. 3.1.)Explanatory successA scientific theory is said to be explanatorily successful just in case it explains natural phenomena that would otherwise be mysterious to us. (See Chap. 3, Sect. 3.1.)

**Fallacious argument** An argument whose premises fail to provide either conclusive or probable support for its conclusion (see also *invalid argument* and *weak argument*). (See Chap. 2, Sect. 2.2.)

**Hasty generalization** A fallacious inductive argument from a sample that is not representative of the general population that is the subject of the conclusion of the argument (because the sample is too small or cherry-picked rather than randomly selected). (See Chap. 2, Sect. 2.2.)

**The historical graveyard of science** The claim that, throughout the history of science, most scientific theories and theoretical posits have been abandoned, discarded, or replaced by new scientific theories and theoretical posits. (See Chap. 5, Sect. 5.1.)

**Inference to the Best Explanation (IBE)** An ampliative (or non-deductive) form of argumentation that proceeds from a phenomenon that requires an explanation to the conclusion that the best explanation for that phenomenon is probably true. (See Chap. 4, Sect. 4.1.)

**Invalid argument** A deductive argument in which the premises purport but fail to provide logically conclusive support for the conclusion. (See Chap. 1, Sect. 1.1.)

**Modus ponens** A form of argument with a conditional premise, a premise that asserts the antecedent of the conditional premise, and a conclusion that asserts the consequent of the conditional premise. That is, "if $A$, then $B$, $A$; therefore, $B$," where $A$ and $B$ stand for statements. *Modus ponens* is a valid form of inference, and so an argument in natural language that takes this logical form is valid. On the other hand, the following logical form is invalid: "if $A$, then $B$, $B$; therefore, $A$." It is known as the fallacy of affirming the consequent. (See Chap. 4, Sect. 4.1.)

**Modus tollens** A form of argument with a conditional premise, a premise that denies the consequent of the conditional premise, and a conclusion that denies the antecedent of the conditional premise. That is, "if $A$, then $B$, not $B$; therefore, not $A$," where $A$ and $B$ stand for statements. *Modus tollens* is a valid form of inference, and so an argument in natural language that takes this logical form is valid. On the other hand, the following logical form is invalid: "if $A$, then $B$, not $A$; therefore, not $B$." It is known as the fallacy of denying the antecedent. (See Chap. 5, Sect. 5.1.)

**Predictive success** A scientific theory is said to be predictively successful just in case it makes predictions that are borne out by observation and experimentation. (See Chap. 3, Sect. 3.1.)

**The Problem of Unconceived Alternatives (PUA)**  The claim that, throughout the history of science, scientists typically occupied an epistemic position in which they could conceive of only a few theories that were well-confirmed by the available evidence, while there were alternative theories that were as well-confirmed by the available evidence as those theories that were accepted by scientists. (See Chap. 5, Sect. 5.4.)

**Scientific realism**  An epistemically positive attitude toward those aspects of scientific theories that are worthy of belief. Scientific realism comes in different varieties, such as Explanationist Realism (see Chap. 3, Sect. 3.1), Entity Realism (see Chap. 3, Sect. 3.4), Structural Realism (see Chap. 3, Sect. 3.5), and Relative Realism (see Chap. 6, Sect. 6.1).

**Weak argument**  A non-deductive (or inductive) argument in which the premises purport but fail to provide probable support for the conclusion. (See Chap. 1, Sect. 1.1.)

# References and Further Readings

Baker, A. (2010). Inference to the best explanation. In F. Russo & J. Williamson (Eds.), *Key terms in logic* (pp. 37–38). London: Continuum.

Bird, A. (2007a). What is scientific progress? *Noûs, 41*(1), 64–89.

Campos, D. G. (2011). On the distinction between Peirce's abduction and Lipton's inference to the best explanation. *Synthese, 180*(3), 419–442.

Chakravartty, A. (2008). What you don't know can't hurt you: Realism and the unconceived. *Philosophical Studies, 137*(1), 149–158.

Chang, H. (2011). The persistence of epistemic objects through scientific change. *Erkenntnis, 75*(3), 413–429.

Douven, I. (2017). Abduction. In E. N. Zalta (Ed.), *The Stanford encyclopedia of philosophy* (Summer 2017 ed.). https://plato.stanford.edu/archives/sum2017/entries/abduction/.

Fahrbach, L. (2011). How the growth of science ends theory change. *Synthese, 180*(2), 139–155.

Frost-Arnold, G. (2011). From the pessimistic induction to semantic antirealism. *Philosophy of Science, 78*(5), 1131–1142.

Godfrey-Smith, P. (2011). Induction, samples, and kinds. In J. K. Campbell, M. O'Rourke, & M. H. Slater (Eds.), *Carving nature at its joins: Topics in contemporary philosophy* (pp. 33–52). Cambridge, MA: The MIT Press.

Harman, G. H. (1965). The inference to the best explanation. *The Philosophical Review, 74*(1), 88–95.

Hesse, M. (1976). Truth and the growth of scientific knowledge. *PSA: Proceedings of the Biennial Meeting of the Philosophy of Science Association, 1976*(2), 261–280.

Hurley, P. J. (2006). *A concise introduction to logic* (9th ed.). Belmont: Wadsworth.

Kitcher, P. (1993). *The advancement of science: Science without legend, objectivity without illusions*. New York: Oxford University Press.

Ladyman, J. (2002). *Understanding philosophy of science*. London: Routledge.

Ladyman, J. (2007). Ontological, epistemological, and methodological positions. In T. Kuipers (Ed.), *General philosophy of science: Focal issues* (pp. 303–376). Amsterdam: Elsevier.

Ladyman, J. (2011). Structural realism versus standard scientific realism: The case of phlogiston and dephlogisticated air. *Synthese, 180*(2), 87–101.

Laudan, L. (1981). A confutation of convergent realism. *Philosophy of Science, 48*(1), 19–49.

Lipton, P. (1993). Is the best good enough? *Proceedings of the Aristotelian Society, 93*(1), 89–104.

Lipton, P. (2005). The Medawar lecture 2004: The truth about science. *Philosophical Transactions of the Royal Society B, 360*(1458), 1259–1269.

Lyons, T. D. (2017). Epistemic selectivity, historical threats, and the non-epistemic tenets of scientific realism. *Synthese, 194*(9), 3203–3219.

Magnus, P. D. (2010). Inductions, red herrings, and the best explanations for the mixed record of science. *The British Journal for the Philosophy of Science, 61*(4), 803–819.

Mizrahi, M. (2013a). The pessimistic induction: A bad argument gone too far. *Synthese, 190*(15), 3209–3226.

Mizrahi, M. (2013b). The argument from underconsideration and relative realism. *International Studies in the Philosophy of Science, 27*(4), 393–407.

Mizrahi, M. (2015). Historical inductions: New cherries, same old cherry-picking. *International Studies in the Philosophy of Science, 29*(2), 129–148.

Mizrahi, M. (2016a). The history of science as a graveyard of theories: A philosophers' myth? *International Studies in the Philosophy of Science, 30*(3), 263–278.

Mizrahi, M. (2016b). Historical inductions, unconceived alternatives, and unconceived objections. *Journal for General Philosophy of Science, 47*(1), 59–68.

Mizrahi, M. (2018). The "positive argument" for constructive empiricism and inference to the best explanation. *Journal for General Philosophy of Science, 3*, 1–6.

Mizrahi, M. (2019). An absurd consequence of Stanford's new induction over the history of science: A reply to Sterpetti. *Axiomathes, 29*(5), 515–527.

Park, S. (2011). A confutation of the pessimistic induction. *Journal for General Philosophy of Science, 42*(1), 75–84.

Park, S. (2019). In defense of realism and selectivism from Lyons's objections. *Foundations of Science, 24*(4), 605–615.

Psillos, S. (2007). The fine structure of inference to the best explanation. *Philosophy and Phenomenological Research, 74*(2), 441–448.

Psillos, S. (2018). Realism and theory change in science. In E. N. Zalta (Ed.), *The Stanford encyclopedia of philosophy* (Summer 2018 ed.) https://plato.stanford.edu/archives/sum2018/entries/realism-theory-change.

Putnam, H. (1975). *Mathematics, matter and method.* New York: Cambridge University Press.

Rosenberg, A. (1980). *Sociobiology and the preemption of social science.* Baltimore: Johns Hopkins University Press.

Roush, S. (2010). Optimism about the pessimistic induction. In P. D. Magnus & J. Busch (Eds.), *New waves in philosophy of science* (pp. 29–58). New York: Palgrave Macmillan.

Rowbottom, D. P. (2019). Extending the argument from unconceived alternatives: Observations, models, predictions, explanations, methods, instruments, experiments, and values. *Synthese, 196*(10), 3947–3959.

Schurz, G. (2019). *Hume's problem solved: The optimality of meta-induction.* Cambridge, MA: The MIT Press.

Stanford, K. P. (2001). Refusing the devil's bargain: What kind of underdetermination should we take seriously? *Philosophy of Science, 68*(S3), S1–S12.

Stanford, K. P. (2006). *Exceeding our grasp: Science, history, and the problem of unconceived alternatives.* New York: Oxford University Press.

Sterpetti, F. (2019). On Mizrahi's argument against Stanford's instrumentalism. *Axiomathes, 29*(2), 103–125.

Tulodziecki, D. (2017). Against selective realism(s). *Philosophy of Science, 84*(5), 996–1007.

van Fraassen, B. C. (1980). *The scientific image.* New York: Oxford University Press.

Vickers, P. (2019). Towards a realistic success-to-truth inference for scientific realism. *Synthese, 196*(2), 571–585.

Weintraub, D. A. (2007). *Is Pluto a planet? A historical journey through the solar system.* Princeton: Princeton University Press.

Wray, B. K. (2008). The argument from underconsideration as grounds for anti-realism: A defence. *International Studies in the Philosophy of Science, 22*(3), 317–326.

Wray, B. K. (2012). Epistemic privilege and the success of science. *Noûs, 46*(3), 375–385.
Wray, B. K. (2018). *Resisting scientific realism*. Cambridge: Cambridge University Press.
Wray, B. K. (2019). Discarded theories: The role of changing interests. *Synthese, 196*(2), 553–569.

# Chapter 6
# Relative Realism: The Best of Both Worlds

**Abstract** This chapter consists of a discussion of my own brand of scientific realism, namely, Relative Realism, first proposed in Mizrahi (Int Stud Phil Sci 27(4):393–407, 2013). Relative Realism provides a middle ground position between scientific realism and antirealism. I take Relative Realism to be a middle ground position between scientific realism and antirealism because it acknowledges the antirealist's point that theory evaluation is comparative (see Chap. 5, Sect. 5.3 on the Underconsideration Argument) while, at the same time, retaining the realist's optimism about science's ability to get *closer* to the truth (that is, to make scientific progress). Unlike other realist positions, Relative Realism is not supported by Inferences to the Best Explanation (IBE), such as the Positive Argument (the so-called "no miracles" argument) for scientific realism (see Chap. 4, Sect. 4.1). Instead, the arguments for Relative Realism are deductive arguments from the comparative nature of theory evaluation and the relative nature of the predictive success of scientific theories, and so they are not subject to the criticisms leveled against many of the realist positions in the scientific realism/antirealism debate that rely on IBE. As such, they are arguments that proceed from premises that both scientific realists and antirealists could accept.

**Keywords** Approximate truth · Comparative evaluation · Comparative realism · Comparative truth · Constructive empiricism · Empirical success · Inference to the best explanation (IBE) · Predictive success · Relative realism · Relative success · Scientific improvement · Scientific progress · Selectionist explanation of success

As we have seen in Chap. 1, the main key to successful argumentation is to argue from premises that one's interlocutor or audience is likely to accept (or at least can be reasonably expected to accept). As far as the arguments for and against scientific realism are concerned, however, our survey of key arguments in Chaps. 4 and 5 has shown that, for the most part, scientific realists tend to argue from premises that antirealists are unlikely to accept, whereas antirealists tend to argue from premises that scientific realists are unlikely to accept. This could explain why some "philosophers have begun to feel that [the scientific realism/antirealism] debate has run its course, or has reached an impasse where neither side is likely ever to be in a position to claim victory," as Brad Wray (2018, p. 1) observes. For example, Arthur Fine

(1986, p. 173) has argued that the "moral of the realism/[antirealism] debate suggests that no reasoned satisfaction is to be had from such a project." I think that it is too early to give up on the scientific realism/antirealism debate. After all, the scientific realism/antirealism debate is a very serious thing, as we have seen in Chap. 1 (see Sect. 1.2), which is why we want just the arguments when it comes to this debate. Accordingly, as long as we argue from premises that both scientific realists and antirealists are likely to accept (or at least can be reasonably expected to accept), we may be able to reach a position that could be acceptable to scientific realists and antirealists alike. In this chapter, then, I aim to do just that. That is to say, I will argue from premises, which both scientific realists and antirealists are likely to accept (or at least can be reasonably expected to accept), as I will attempt to show, for a position that could be attractive to both scientific realists and antirealists.

Accordingly, the position I will argue for in this chapter, namely, "Relative Realism," which was first proposed in Mizrahi (2013),[1] provides a middle ground position between scientific realism and antirealism. I take Relative Realism to be a middle ground position between scientific realism and antirealism because it is inferred from premises that both scientific realists and antirealists are likely to accept (or at least can be reasonably expected to accept). More specifically, Relative Realism acknowledges the antirealist's point that theory evaluation is comparative (see Chap. 5, Sect. 5.3 on the Underconsideration Argument) while, at the same time, retaining the scientific realist's optimism about science's ability to get *closer* to the truth (that is, to make scientific progress). Unlike other realist positions, Relative Realism is not supported by Inferences to the Best Explanation (IBE), such as the Positive Argument (the so-called "no miracles" argument) for scientific realism (see Chap. 4, Sect. 4.1). Instead, the arguments for Relative Realism are deductive arguments from the comparative nature of theory evaluation and the relative nature of the predictive success of scientific theories, and so they are not subject to the criticisms leveled against many of the realist positions in the scientific realism/antirealism debate that rely on IBE. As such, these arguments for Relative Realism are arguments that proceed from premises that both scientific realists and antirealists could accept.

## 6.1  Approximate Truth Versus Comparative Truth

As we have seen in Chap. 2, one of the central questions of the scientific realism/antirealism debate in contemporary philosophy of science is this: "Do we have adequate grounds for believing that our theories are true or approximately true with respect to what they say about unobservable entities and processes?" (Wray 2018, p. 1) According to Relative Realism, we have adequate grounds for believing that,

---

[1] Some text from Mizrahi (2013) is reused in this chapter with permission from Taylor and Francis (license number 4793050820383).

from a set of competing scientific theories, the more empirically successful theory is *comparatively true*, that is, closer to the truth relative to its competitors in the set, rather than approximately true. To understand Relative Realism, then, we need to revisit the notion of approximate truth from Chap. 2. As we have seen in Chap. 2 (see Sect. 2.1), scientific realists and some antirealists agree that, strictly speaking, scientific theories, even the best ones, are not entirely true.[2] The fact that a scientific theory is not entirely true, however, does not necessarily mean that it is completely false. In other words, a scientific theory is not a monolithic whole. Rather, a scientific theory can have some false parts and some true parts (Kitcher 2002, p. 388). In that sense, then, some scientific realists want to talk about a scientific theory as being approximately true in terms of having some true parts and some false parts.[3] Strictly speaking, however, "only propositions can be true or false" (Kvanvig 2003, p. 191), and since a scientific theory is not a single or an individual proposition (at the very least, a scientific theory is a set of propositions), a scientific theory cannot be said to be true or false (Kitcher 1993, p. 118).

By way of illustration, consider the following example, which is adapted from Jarrett Leplin (1997, p. 133). Suppose that there is a power outage in my house. Upon looking outside my window, I see a utility truck parked nearby and some workers digging in the yard. Since I made a call to the phone company earlier about a problem with my phone line, I infer that technicians from the telephone company, who have responded to my earlier call, inadvertently cut the power line to my house. Unbeknownst to me, however, it is not technicians from the telephone company who have cut the power line but rather technicians from the cable company whom I had not expected. Now, if we take this "theory," that is, that there is a power outage in my house because technicians from the telephone company have inadvertently cut the power line to my house, as a monolithic whole, then it is strictly false. However, this theory involves several claims, some of which are true, whereas others are false. On the one hand, it is not the case that telephone technicians working in the backyard have inadvertently cut the power line. On the other hand, it is true that technicians working in the backyard have inadvertently cut the power line. I may not know the truth, the whole truth, and nothing but the truth about this state of affairs. But I do know some parts about it, and those parts are themselves true.

Consider an example from the history of science next. In his *An Inquiry into the Causes and Effects of the Variolae Vaccinae*, Edward Jenner argues that cowpox originated as grease, a disease common in horses. He claims that it was transmitted to cows when horse handlers helped with milking on occasion. In addition, Jenner (1800, p. 7) claims not only that cowpox protected against smallpox but also that "what renders the Cow Pox virus so extremely singular, is, that the person who has been thus affected is for ever after secure from the infection of the Small Pox." Now, if we take Jenner's entire *Inquiry* as his "theory," then it is strictly false as a

---

[2] As Anjan Chakravartty (2017) observes, "it is widely held, not least by realists, that even many of our best scientific theories are likely false, strictly speaking."

[3] As we have seen in Chap. 2, however, providing a precise formal (or even an informal) definition of approximate truth has proved to be very difficult. See Chap. 2, Sect. 2.1.

monolithic whole. He was wrong about grease being the origin of cowpox. He mistakenly took horsepox for grease, and there was no intermediate passage through cows, either. Even though he got some things wrong, he was right about others. His hypothesis, properly construed, is correct. While it is not the case that vaccination provides lifelong protection, as Jenner thought, it is the case that repeated vaccination, properly done, contributes to the control of smallpox. Indeed, Jenner paved the way for this knowledge, and the know-how for the selection of correct material for vaccination, with his distinction between true and spurious cowpox. Nowadays, pseudocowpox (Milker's nodes) is recognized as a type of spurious cowpox (Baxby 1999). According to the World Health Organization, "Publication of the *Inquiry* and the subsequent energetic promulgation by Jenner of the idea of vaccination with a virus other than variola virus constituted a watershed in the control of smallpox, for which he more than anyone else deserves the credit" (Fenner et al. 1988, p. 264).

Another example from the history of science is Paul Ehrlich's side-chain theory of antibody formation. Ehrlich proposed that harmful compounds could mimic nutrients for which cells express specific receptors. However, he considered these receptors to be on all cell types. He also did not realize that there are specialized producer cells, such as B-lymphocytes. He thought of the entire spectrum of receptors as a single cell because he considered their main task as the uptake of different nutrients. These are the parts of Ehrlich's side-chain theory that turned out to be incorrect. It does not follow, however, that the entire theory is wrong. Despite these errors, the theory is based on a correct principle, which is that "specific receptors on cells interact with foreign material in a highly specific way, and this triggers their increased production and release from the cell surface so that they can inactivate foreign material as antibodies" (Kaufmann 2008, p. 707).

These examples illustrate how scientific theories are not monoliths that can be said to be true or false as a whole in the same way that single or individual proportions can be said to be true or false. If this is correct, then it seems that we need to abandon talk of whole theories as being true or false. Instead, we should talk about theoretical propositions or claims as being true or false individually, but not about whole theories as being true or false. Indeed, antirealists seem to acknowledge this point. For instance, Brad Wray (2008, p. 323) writes, "For the sake of clarity, let me call $H_1$ the Tychonic hypothesis, rather than the Tychonic theory. After all, *the Tychonic theory includes an array of other claims*" (emphasis added). Likewise, Wray (2010, p. 6) also writes, "our theories, consisting of many *theoretical claims*, that is, a conjunction of numerous theoretical claims, are most likely false" (emphasis in original). Since scientific theories consist of many theoretical claims, which means that they are not single or individual propositions, but "only propositions can be true or false" (Kvanvig 2003, p. 191), it follows that scientific theories cannot be said to be true or false, strictly speaking.

If theoretical claims (as single or individual propositions), rather than whole theories, are the sort of things that can be said to be true or false, then we need to distinguish between truth and approximate truth along the following lines. Approximate truth, which is said of theories, is not like truth, which is said of propositions (or theoretical claims), insofar as the former is *relative*, whereas the latter is *categorical*. This is because the only way to make judgements about the approximate truth of scientific theories is by testing those theories. As we have seen in Chap. 5 (see Sect. 5.3), when scientists test theories, they compare a few competing theories. In other words, theory evaluation is comparative. As Brad Wray (2018, p. 4) puts it, "evaluations of competing theories are comparative in nature." Accordingly, from a set of competing theories, if one theory $T$ passes the tests (that is, its predictions are borne out by observational and experimental testing), then that is a reason to believe that $T$ is *closer* to that truth than its competitors are. If this is correct, then to say that $T$ is approximately true is to say that $T$ is *closer* to the truth than its competitors are. In that sense, approximate truth is *relative*, that is, it is a comparative relation between competing theories.

To sum up, then, truth is a property of propositions (or theoretical claims), since only propositions (or theoretical claims) can be categorically true (or false), whereas approximate truth is a comparative relation between competing theories, since a theory can be closer to the truth only relative to its competitors. Some might object, however, that scientific theories, expressed as sets of propositions, are simply conjunctions, and conjunctions are categorically truth or false (cf. Wray 2010, p. 6). For example, one could argue that the germ theory of disease is just a conjunction of the following theoretical claims (Tulodziecki 2016, pp. 272–273):

- Diseases are biological;
- Diseases are the result of specific pathogenic microorganisms, which are called "germs";
- Germs are fixed;
- Different diseases are associated with different germs;
- Transmission occurs from a diseased host to another individual.

The conjunction of these theoretical claims, then, is the theory called "the germ theory of disease," or so the objection goes. And one could add further theoretical claims to this long conjunction. For example, perhaps one would like to add Robert Koch's postulates, according to which, in order to establish a causal connection between a specific microorganism and a specific disease, one must be able to demonstrate the following empirically (Lim 2003, p. 8):

- The specific microorganism is present in all cases of the specific disease.
- The specific microorganism can be isolated from the diseased host and grown in pure culture.
- The isolated microorganism when inoculated into a healthy, susceptible host, will cause the same disease seen in the original host.
- The microorganism can be reisolated in a pure culture from the experimental infection.

At any rate, the objection is that the germ theory of disease is just a conjunction of theoretical claims, no matter how long or complex the conjunction turns out to be.

In reply, I would argue that the distinction between truth and approximate truth is analogous to the logical distinction between truth and validity. In logic, and as we have done in this book as well, we distinguish between propositions or statements, which can serve as premises and conclusions in arguments, and arguments as sets of propositions or statements in which some (at least one proposition or statement called a *premise*) purport to provide logical support (either conclusive, as in deductive arguments, or probable, as in non-deductive or inductive arguments) for another proposition or statement (namely, the *conclusion*). Truth is a categorical property of premises and conclusions, but not arguments, whereas validity is a relational property of deductive arguments, but not premises or conclusions. In other words, we talk of deductive arguments as being valid or invalid, but not true or false, as we have been doing in this book. This is so, even though, in principle, a deductive argument can be expressed as a conditional (that is, if the premises are true, then the conclusion cannot be false), which is categorically true or false. In logic, we reserve the terms 'true' and 'false' to premises and conclusions, and the terms 'valid' and 'invalid' to deductive arguments, in order to capture the difference between truth as a categorical property of propositions or statements and validity as a relation between propositions or statements (more specifically, a relation between premises and a conclusion). Similarly, I propose, we should reserve the term 'true' for theoretical claims, which are single or individual propositions that can be categorically true or false, and the term 'approximately true' for theories, since it expresses a comparative relation between competing theories, even though, in principle, theories can be expressed as conjunctions of theoretical claims.

Since I am introducing a new way of talking about approximate truth, namely, as a comparative relation between competing theories rather than a categorical property of theoretical claims, I will only talk of *comparative truth* rather than approximate truth from now on. Accordingly, to say that $T_1$ is comparatively true is to say that $T_1$ is closer to the truth than its competitors, $T_2$, $T_3$, ..., $T_n$, since comparative truth is a comparative relation between competing theories. To address some obvious worries about this notion of comparative truth, note that, if $T_1$ is closer to the truth than its competitors, $T_2$ and $T_3$, then $T_2$ can still be closer to the truth than $T_3$, and thus comparatively true relative to $T_3$. Moreover, if $T_1$ has no competitors, then $T_1$ cannot be comparatively true, since being comparatively true consists in being closer to the truth relative to competing theories. If there are no competing theories, then a theory cannot be comparatively true. Finally, if an evaluation of competing theories $T_1$, $T_2$, and $T_3$ yields the verdict that $T_1$ is closer to the truth than $T_2$, and $T_3$, then $T_1$ is comparatively true, no matter how far from the truth $T_1$ is absolutely speaking. This is so because theory evaluation is comparative. That is to say, "evaluations of competing theories are comparative in nature" (Wray 2018, p. 4). In that sense, comparative truth is a comparative relation between competing theories; a scientific theory is comparatively true based on the way it outperforms its competitors in observational and experimental testing.

With this understanding of the key notion of Relative Realism, namely, *comparative truth*, as a comparative relation between competing theories, as opposed to a categorical property of theoretical statements, we are now in a position to consider the arguments for Relative Realism. As we will see in what follows, the arguments for Relative Realism are not instances of Inference to the Best Explanation (IBE) like the Positive Argument (the so-called "no miracles" argument) for scientific realism (see Chap. 4, Sect. 4.1). Instead, they are deductive arguments from the comparative nature of theory evaluation and the relative nature of the predictive success of scientific theories, and so they are not subject to the criticisms leveled against many of the realist positions in the scientific realism/antirealism debate that rely on IBE (see Chap. 4, Sect. 4.1). As such, they are arguments that proceed from premises that both scientific realists and antirealists could accept (or at least can be reasonably expected to accept).

## 6.2   The Kuhnian Argument from the Illusiveness of the Truth of Whole Theories

As we have seen in Sect. 6.1, *truth* is a property of theoretical claims, whereas *comparative truth* is a comparative relation between competing theories, which have theoretical claims as their parts. The former is categorical, whereas the latter is relative. Accordingly, to say that $T_1$ is comparatively true is to say that $T_1$ is closer to the truth than its competitors, $T_2$, $T_3$, ..., $T_n$. According to Relative Realism, then, we have good reasons to believe that, from a set of competing scientific theories, the more empirically successful theory is *comparatively true*, that is, closer to the truth relative to its competitors in the set (Mizrahi 2013). This position in the scientific realism/antirealism debate in contemporary philosophy of science is motivated by an argument that can be reconstructed from the Postscript to the second edition of Thomas Kuhn's seminal book, *The Structure of Scientific Revolutions* (1962/1996, pp. 206–207).

> Kuhn, Thomas S. 1962/1996. *The structure of scientific revolutions*. Chicago: The University of Chicago Press.

> Perhaps there is some other way of salvaging the notion of 'truth' for application to *whole theories*, but this one will not do. There is, I think, no *theory-independent way* to reconstruct phrases like 'really there'; the notion of a *match* between the ontology of a theory and its "real" counterpart in nature now seems to me *illusive in principle*. Besides, as a historian, I am impressed with the implausability of the view. I do not doubt, for example, that Newton's mechanics improves on Aristotle's and that Einstein's improves on Newton's as instruments for puzzle-solving. But I can see in their succession no coherent direction of ontological development. On the contrary, in some important respects, though by no means in all, Einstein's general theory of relativity is closer to Aristotle's than either of them is to Newton's. Though the temptation to describe that position as relativistic is understandable, the description seems to me wrong. Conversely, if the position be *relativism*, I cannot see that the relativist loses anything needed to account for the nature and development of the sciences (emphasis added).

As we have seen in Sect. 6.1, the relative realist would agree with Kuhn that it makes no sense to talk about whole theories as being true (or false). This is because a scientific theory consists of several theoretical claims. Since, strictly speaking, "only propositions can be true or false" (Kvanvig 2003, p. 191), and a scientific theory consists of more than a single proposition or theoretical claim, it follows that a scientific theory cannot be said to be true (or false), strictly speaking. Instead, the relative realist would argue, we should talk about *comparative truth*. To say that $T_1$ is comparatively true is to say that $T_1$ is closer to the truth than its competitors, $T_2$, $T_3$, ..., $T_n$, since comparative truth is a relation between competing theories.

Now, the argument Kuhn seems to be hinting at in the quoted passage above, which is of interest for our present purposes, is an argument for the conclusion that "the notion of a match between the ontology of a theory and its 'real' counterpart in nature [is] illusive in principle" (Kuhn 1962/1996, p. 206). This conclusion is supposed to follow from the fact that there is "no theory-independent way to reconstruct phrases like 'really there'" (Kuhn 1962/1996, p. 206). In other words, if we would like to say that a scientific theory, $T$, is true as a whole just in case its theoretical claims, that is, claims about the unobservable entities, processes, or events it posits in order to explain and predict some observable phenomenon, $P$, match with "what is really the case," then we need to know that $T$ and "what is really the case" match. That is to say, if the truth of $T$ (for example, the three-dimensional, double-helical model of DNA) is a matter of its matching the facts about some domain, $D$ (for example, the deoxyribonucleic acid or DNA molecule), then knowing that $T$ is true is a matter of knowing that there is a match between $T$ and the facts about $D$. Now, in order to know that there is a match between $T$ and the facts about $D$, we need to be able to consider $T$ and the facts about $D$ independently and then determine whether they match or not. But, Kuhn would argue, there is no way we could do that, for access to the facts about $D$ is always mediated by scientific theories, that is, scientific theories just like $T$. Since we do not have unmediated access to the facts about $D$, that is, we cannot consider the facts about $D$ independently of some scientific theory just like $T$, it follows that we cannot know that $T$ is true as a whole. In that sense, Kuhn would argue, the truth of whole theories is illusive in principle. This Kuhnian argument can be stated in canonical (or standard) form as follows:

(P1) If the truth of a whole theory, $T$, is a matter of its matching a scientific fact, $F$, then in order to know that $T$ is true, we need to know that $T$ and $F$ match.
(P2) To know that $T$ and $F$ match, we need to examine $T$ and $F$ independently and then determine that they match.
(P3) But we cannot examine $T$ and $F$ independently, since access to $F$ is always mediated by scientific theories, such as $T$.

Therefore,

(C1) We cannot know that $T$ and $F$ match. [from (P2) & (P3)].

Therefore,

(C2) It is not the case that the truth of a whole theory, $T$, is a matter of its matching a scientific fact, $F$. [from (P1) & (C1)].

This line of reasoning is valid. That is to say, at each step, the premises provide logically conclusive support for the conclusion that follows from those premises. The next question, then, is whether the premises are actually true. Is the Kuhnian Argument from the Illusive Truth of Whole Theories a sound argument?

The key premise of the Kuhnian Argument from the Illusive Truth of Whole Theories seems to be (P3). As we have seen in Chap. 4 (see Sect. 4.4), the physical means by which scientists study domains in nature are *theoretically laden*. That is to say, the scientific instruments and techniques that scientists use in their investigations of the natural world are based on theoretical knowledge from physics, optics, and other scientific fields of study.[4] If the scientific instruments and techniques that are used to study scientific facts about theoretical posits, that is, unobservable entities, processes, and events, such as genes and genetic mutations, are dependent on theoretical knowledge, then there is no independent access to scientific facts about those theoretical posits that is not mediated by theories, just as (P3) of the Kuhnian Argument from the Illusive Truth of Whole Theories states. As Michela Massimi (2004, p. 39) puts it, "evidence in favour of [unobservable entities, such as] colored quarks [comes] from extraordinarily theory-loaded experiments." In other words, scientists gain access to what is "really there" only by means of observation aided by scientific instruments, such as microscopes and telescopes, which depend on theoretical knowledge from various scientific fields to work as they do. In that respect, our access to scientific facts is mediated by the theoretical knowledge that we use to build the very scientific instruments that are supposed to give us theoretical knowledge of scientific facts in the first place.[5]

Now, the relative realist would have no problem with (P3) of the Kuhnian Argument from the Illusive Truth of Whole Theories. That is to say, the relative realist can accept that there is no direct, theoretically unmediated access to the sort of domains in nature that scientists study using their scientific instruments of observations and methods of experimentation. Moreover, just as Kuhn does, the relative realist would also find talk of the truth of whole theories rather nonsensical. For the relative realist, then, the lesson to draw from the Kuhnian Argument from the Illusive Truth of Whole Theories is that there is no direct, theory-independent way to compare our scientific theories with the facts in order to see that they match. So we should give up on the notion that our current best scientific theories are likely or approximately true in the sense "of a *match* between the ontology of a theory and its 'real' counterpart in nature" (Kuhn 1962/1996, p. 206). Such theoretical knowledge is illusive in principle, as Kuhn would argue. But that does not mean that we should give up on theoretical knowledge altogether. The sort of theoretical knowledge we can have, the relative realist would argue, is *comparative* knowledge. That is to say, while there is no direct, theory-independent way to compare our scientific theories with the facts in order to see if they match, there is a way to *compare our scientific theories to other scientific theories* in order to see which is closer to the truth. In

---

[4]According to Hasok Chang (2004, p. 52), "it is now widely agreed that observations are indeed affected by the theories we hold, thanks to the well-known persuasive arguments to that effect by Thomas Kuhn, Paul Feyerabend, Marry Hesse, Norwood Russell Hanson and even Karl Popper, as well as various empirical psychologists." See also Bogen (2020).

[5]According to Stathis Psillos (1999, p. 76), "That the methods by which scientists derive and test theoretical predictions are theory-laden is undisputed. [...] All aspects of scientific methodology are deeply theory-informed and theory-laden."

other words, scientific truth is relative insofar as a scientific theory can be *comparatively true*, that is, closer to the truth *relative to its competitors*. If this is correct, then scientists do not compare their theories to the world, as Kuhn would argue. Scientists do not even compare competing theories with each other *and the world*. Instead, scientists compare competing theories with each other only.

Accordingly, unlike Kuhn, who seems to have shied away from the notion of relativity, the relative realist embraces that notion with respect to scientific theories. For, by embracing the relativity of scientific theories, the relative realist does not lose "anything needed to account for the nature and development of the sciences" (Kuhn 1962/1996, p. 207), as Kuhn himself puts it. In particular, the relative realist has the comparative nature of theory evaluation in science and the relative nature of the predictive success of scientific theories. Since many scientific realists (and some antirealists) have opposed relativism, however, it is important to clarify the sort of relativity that the relative realist says is an integral part of science. As Thomas Nickles (2020) puts it:

> For many philosophers of science, relativism is the big bugaboo that must be defeated at all costs. For them, any view that leads to even a moderate relativism is thereby reduced to absurdity. Historicist philosophers have insisted on relativity to historical context but, with few exceptions, have made a sharp distinction between *relativity* and outright *relativism* (emphasis in original).

Indeed, the relative realist would insist on drawing a sharp distinction between *relativity* and *relativism*. The relative realist defends the former, rather than the latter, as applied to theoretical knowledge of unobservables in science. That is to say, in relativistic physics, the relativity principle entails that "it is impossible for an observer in an inertial system to determine absolute rest or absolute motion for the system" (Topper 2013, p. 45). Similarly, the relative realist would argue, it is impossible for scientists to determine the *absolute truth* of scientific theories, as the Kuhnian Argument from the Illusive Truth of Whole Theories purports to show. Instead, scientists can only determine the *comparative truth* of theories relative to some frames of reference. Those frames of reference are the other theories with which scientists compare any given theory. This sort of relativity is different from the "big bugaboo" relativism that many philosophers of science think "must be defeated at all costs" (Nickles 2020) because it does not imply that "anything goes." That is to say, according to the "big bugaboo" relativism that many philosophers of science reject, all claims to theoretical knowledge in science are equally legitimate because there are no fixed standards by which to judge those claims to theoretical knowledge, and so "anything goes"; your claim to theoretical knowledge is as good as mine. In other words, according to "big bugaboo" relativism, all scientific theories are "born equal," so to speak, and thus no theory can be said to be better than another because there are no fixed standards by which to make such comparative judgments about scientific theories. According to Relative Realism, however, there are fixed standards for evaluating scientific theories, such as observational and experimental testability, as well as other theoretical virtues (see Chap. 4, Sect. 4.1). Consequently, Relative Realism does not imply that "anything goes." As long as scientists can evaluate scientific theories and rank them in order from better to worse, which the relative realist claims they can, it is not the case that "anything goes" in science. Such ranking of theories, however, is relativized insofar as scientific theories can only be judged better or worse relative to their competitors. In other words, in science, theory evaluation (that is, the relativized ranking of competing theories) is inherently comparative, and thus it can only yield comparative knowledge.[6]

In that respect, perhaps "Relativistic Realism" would have been a better name than "Relative Realism." However, Nicholas Rescher (1990) has already defended a view he calls "Relativistic Realism." According to Rescher (1990, p. 102). Relativistic Realism is.

> a realism that is relativistic in that its insistence on the multi-faceted nature of the real means that any science will reflect its deviser's particular 'slant' on reality (in line with the investigator-characteristic modes of interaction with nature). On such a view, knowledge of reality is always (in some crucial respect) cast in terms of reference that reflect its possessor's cognitive proceedings. There is, no doubt, a mind-independent reality, but *cognitive access to it is always mind-conditioned* (emphasis added).

The relative realist would agree that our cognitive access to a mind-independent reality is always mediated by our theoretical frameworks, as we have seen from the Kuhnian Argument from the Illusive Truth of Whole Theories. However, the relative realist would not say that scientific knowledge is relative to the knower. Rather, the very act of producing scientific knowledge is comparative, that is, scientists produce scientific knowledge by comparing competing theories with each other, and so a scientific theory can only be *comparatively true*, that is, closer to the truth *relative to its competitors*.

Alternatively, perhaps "Comparative Realism" would have been a better name than "Relative Realism." However, Theo Kuipers (2019) has defended a view he calls "Comparative Realism." I will compare Comparative Realism and Relative Realism in Sect. 6.5. There are important differences between these two positions. In the meantime, with comparative theory evaluation and relative success in hand, the relative realist has everything that is needed to have a middle ground position between scientific realism and antirealism, which is a position that meets both the scientific realist and the antirealist where they are. With this preliminary motivation for Relative Realism from the Kuhnian Argument from the Illusive Truth of Whole Theories in hand, then, I now turn to making the arguments for Relative Realism: the first argument from the comparative nature of theory evaluation in science and the second argument from the relative nature of the predictive success of scientific theories.

---

[6]Natalie Alana Ashton (2020, p. 83) makes a similar point about Michela Massimi's (2018) Perspectival Realism. Perspectival Realism, which stems from the work of Ronald Giere (2006), is the view that there is a single way the world is (cf. the metaphysical stance/thesis of scientific realism in Chap. 2, Sect. 2.1), but there are many, equally legitimate ways in which one might come to know about that world. In that respect, Perspectival Realism seems more like Nicholas Rescher's Relativistic Realism (see the main text) than Relative Realism, since truth or knowledge is relativized to a perspective taken by a knower.

## 6.3   The Argument from the Comparative Evaluation of Theories

As we have seen in Sect. 6.1, *truth* is a property of theoretical claims, whereas *comparative truth* is a relation between competing theories, which have theoretical claims as their parts. The former is categorical, whereas the latter is relative. Accordingly, to say that $T_1$ is comparatively true is to say that $T_1$ is closer to the truth than its competitors, $T_2, T_3, ..., T_n$. As we have seen in Sect. 6.2, the notion of comparative truth provides the conceptual framework for Relative Realism, according to which we have good reasons to believe that, from a set of competing scientific theories, the more empirically successful theory is *comparatively true*, that is, closer to the truth relative to its competitors in the set (Mizrahi 2013). As I see it, Relative Realism is a middle ground position between scientific realism and antirealism. This is because Relative Realism acknowledges the antirealist's point that theory evaluation is comparative while, at the same time, retaining the realist's optimism about science's ability to get *closer* to the truth (that is, to make scientific progress).

As we have seen in Sect. 6.2, Relative Realism is motivated by an argument that can be reconstructed from the Postscript to the second edition of Thomas Kuhn's seminal book, *The Structure of Scientific Revolutions* (1962/1996, pp. 206–207). The key arguments for Relative Realism, however, are arguments from the comparative nature of theory evaluation and the relative nature of the predictive success of scientific theories. So let us turn to these arguments now. The first argument for Relative Realism is the Argument from the Comparative Evaluation of Theories.

As we have seen in Sect. 6.1, antirealists like Brad Wray (2018, p. 4) argue that "evaluations of competing theories are comparative in nature." If we can evaluate theories comparatively, then we can make comparative judgments about the relative superiority of one theory over its competitors. In other words, we can say that one theory is better, that is, closer to the truth, than its competitors (that is, comparatively true) based on a comparative evaluation of these competing theories. In that sense, an argument for Relative Realism can be made from the comparative nature of theory evaluation. This argument can be stated in canonical (or standard) form as follows:

(P1) In evaluating theories, scientists rank the competitors comparatively.
(P2) If scientists rank competing theories comparatively, then they are justified in believing comparative judgements (that is, $T_1$ is closer to the truth than competitors $T_2, T_3, ..., T_n$), rather than absolute judgements (that is, $T_1$ is likely true), about competing theories.

Therefore,

(C) Scientists are justified in believing comparative judgements (that is, $T_1$ is closer to the truth than competitors $T_2, T_3, ..., T_n$), rather than absolute judgements (that is, $T_1$ is likely true), about competing theories.

As stated, this Argument from the Comparative Evaluation of Theories is a deductive argument (in particular, a *modus ponens*: If $A$, then $B$, $A$; therefore, $B$) for Relative Realism. Since the premises of this argument, namely, (P1) and (P2), successfully provide logically conclusive support for the conclusion, namely, (C), this argument can be said to be valid. The next question, then, is whether the premises are in fact true. Is the Argument from the Comparative Evaluation of Theories a sound argument for Relative Realism?

As we have seen in Sect. 6.1, to make comparative judgements about competing theories is to say that a theory is comparatively true or that it is closer to the truth relative to its competitors (that is, $T_1$ is closer to the truth than its competitors $T_2$, $T_3$, …, $T_n$). The upshot of the Argument from the Comparative Evaluation of Theories, then, is that theory evaluation can give us reasons to believe that a theory is comparatively true, that is, that $T_1$ is *closer* to the truth than competitors $T_2$, $T_3$, …, $T_n$, but it cannot give us reasons to believe that a theory is *closest* to the truth (that is, that $T_1$ is likely true). To illustrate, consider the following logical space of possible theories relative to the truth:

$T_1$   $T_2$   $T_3$   $T_4$   $T_5$   $T_6$   $T_7$   $T_8$   TRUTH

If scientists evaluate $T_1$, $T_2$, and $T_3$ by observational and experimental testing, they could reasonably make the comparative judgement that $T_3$ is comparatively true relative to competitors $T_2$ and $T_1$ when $T_3$ outperforms $T_1$ and $T_2$ in such tests. However, a theory can be closer to the truth relative to its competitors but still be quite far off from the truth, given the logical space of possible theories. Theory evaluation cannot tell us which theory is *close* or *closest* to the truth, unless we have independent reasons to believe that the theories we are testing are those that are *close* or *closest* to the truth (that is, $T_7$ and $T_8$). Since we do not have independent reasons to believe that, however, we cannot reasonably claim that the theories we have tested are *close* or *closest* to the truth (that is, likely true), although we can reasonably claim that one of them is *closer* to the truth than its competitors (that is, that a theory is comparatively true). In other words, theory evaluation can tell us which theory among competing theories is closer to the truth (for example, that $T_3$ is closer to the truth than $T_1$ and $T_2$). However, theory evaluation cannot tell us which theory among competing theories is close or closest to the truth.

Another way to illustrate this crucial point about Relative Realism is to consider the following sets of competing theories:

Lot 1 $\{T_1, T_2, T_3, T_4\}$
Lot 2 $\{T_5, T_6, T_7, T_8\}$

Theory evaluation can justify the belief that $T_4$ is comparatively true, which is elliptical for $T_4$ is closer to the truth than competitors $T_1$, $T_2$, and $T_3$. But theory evaluation cannot justify the belief that $T_4$ is likely true (or close/closest to the truth). Similarly, theory evaluation can justify the belief that $T_8$ is comparatively true, which is elliptical for $T_8$ is closer to the truth relative to competitors $T_5$, $T_6$, and $T_7$. But theory evaluation cannot justify the belief that $T_4$ is likely true (or close/closest to the truth). Nor can theory evaluation justify the belief that $T_8$ is likely true (or close/closest to the truth). This is because we do not know if (or have good reasons to believe whether) we are testing theories from Lot 1 or Lot 2. In other words, we have no way of knowing (or justifiably believing) the position of the theories we are testing in the logical space of possible theories relative to the truth (Cf. Park 2017a).

As we have seen in Sect. 6.1, antirealists would accept (P1) of the Argument from the Comparative Evaluation of Theories, since they would agree that "evaluations of competing theories are comparative in nature" (Wray 2018, p. 4). In fact, Brad Wray uses (P1) of the Argument from the Comparative Evaluation of Theories as a premise in his own version of the underconsideration argument against scientific realism (see Chap. 5, Sect. 5.3). According to Wray (2012, p. 377), the antirealist argues as follows:

P1. In evaluating theories scientists merely rank the competitors comparatively.
P2. Scientists are not epistemically privileged, that is, they are not especially prone to develop theories that are true with respect to what they say about unobservable entities and processes.

C. Hence, we have little reason to believe that the theory that is judged to be superior is likely true.

Like Wray, the relative realist asserts that theory evaluation is comparative. That is to say, when scientists evaluate theories, they rank competing theories comparatively. Antirealists like Wray take this "ranking premise," as Peter Lipton (1993, p. 89) calls it, to be a reason to believe that we have no good reasons to believe that the theory that is judged to be superior is likely true. The relative realist would agree with that. Indeed, we do not have good reasons to believe that the theory that is judged to be superior to its competitors is *likely true*. But, the relative realist would argue, we do have good reasons to believe that the theory that is judged to be superior to its competitors is *comparatively true*, that is, closer to the truth than its competitors. This is precisely because theory evaluation is comparative, which means that such theory testing yields comparative warrant. As Lipton (1993, p. 89) puts it, theory "testing enables scientists to say which of the competing theories they have generated is likeliest to be correct, but does not itself reveal how likely the likeliest theory is." In other words, comparative theory testing enables scientists to judge which theory of a set of competing theories is closer to the truth relative to its competitors in the set, but it does not enable scientists to judge which theory is close or closest to the truth, absolutely speaking.

It is evident, then, that antirealists would accept (P1) of the Argument from the Comparative Evaluation of Theories. I think that antirealists would also accept (P2). For they would admit that theory evaluation entails that one theory could be judged to be superior to another theory. For instance, Larry Laudan (2004, p. 17) grants that "a theory is, all else being equal, *better* if it can explain or predict facts from different domains or if it can show its rivals to be limiting cases" (emphasis added). Of course, to say that a theory $T_1$ is better than a competing theory $T_2$ is to make a comparative judgment about two theories. The relative realist would argue that such comparative judgments are justified by the comparative nature of theory evaluation itself, and it seems that Laudan would agree with that as well.

Along the same lines, Wray (2018, p. 45) admits that "the superior theory is *more likely true than* the competitor theories with which it is compared" (emphasis added). Like Laudan, then, it appears that Wray would also grant that comparative judgments, such as $T_1$ is closer to the truth than competitors $T_2$, $T_3$, ..., $T_n$, can be justified by theory evaluation. If so, then Wray would have to grant the key tenet of Relative Realism, which is that comparative judgments, such as $T_1$ is closer to the truth than competitors $T_2$, $T_3$, ..., $T_n$, are justified by the comparative nature of theory evaluation. By its very nature, which is *comparative*, theory evaluation allows us to say which theory among a set of competing theories is *comparatively true*, but it does not allow us to say which theory is absolutely or likely true.

Moreover, Wray (2018, p. 49) also grants that scientists are reliable "with respect to their judgments of those features of theories that they can ascertain directly, like predictive accuracy." If their judgments of which theory among several competing theories is superior in terms of predictive accuracy are reliable, then scientists are in a position to judge that, among those competing theories, the superior theory is comparatively true, that is, the superior theory is closer to the truth relative to its competitors. All of this seems to suggest that antirealists would accept the premises of the Argument from the Comparative Evaluation of Theories. Now, if the Argument from the Comparative Evaluation of Theories is valid, and its premises are true, or at least acceptable to antirealists, then it can be said to be a sound argument for Relative Realism.

Like antirealists, scientific realists should also accept the key premise of the Argument from the Comparative Evaluation of Theories, namely, (P1). For scientific realists would grant that theory choice in science consists of comparing and choosing between competing theories. For instance, according to Alan Musgrave (2017, p. 71), "As everybody also knows, science is empirical--scientists use observation and experiment to try to *decide between the competing theories* that they propose" (emphasis added). Likewise, according to Stathis Psillos (1999, p. 76), "Scientists use accepted background theories in order to form their expectations, to choose the relevant methods for theory-testing, to devise experimental set-ups, to calibrate instruments, to assess the experimental evidence, *to choose among competing theories*, to assess newly suggested hypotheses, etc." (emphasis added). If theory choice in science is inherently comparative, that is, consisting of "choosing between competing theories" (Cordero 2000, p. 191), then such comparative theory evaluation, by its very nature, entitles scientists to make comparative judgments about the competing theories they are evaluating, just as (P2) of the Argument from the Comparative Evaluation of Theories states. As Lipton (1993, p. 89) puts it, theory "testing enables scientists to say which of the competing theories they have generated is likeliest to be correct, but does not itself reveal how likely the likeliest theory is." All of this seems to suggest that, like antirealists, scientific realists would also accept the premises of the Argument from the Comparative Evaluation of Theories. Now, if the Argument from the Comparative Evaluation of Theories is valid, and its premises are true, or at least acceptable to scientific realists, then it can be said to be a sound argument for Relative Realism.

## 6.4   The Argument from the Relative Success of Theories

As we have seen in Sect. 6.3, from the comparative nature of theory evaluation in science, it follows that scientists are justified in believing comparative judgements about the comparative truth of competing theories (that is, $T_1$ is closer to the truth relative to competitors $T_2$, $T_3$, ..., $T_n$), rather than absolute judgements about the truth of competing theories (that is, $T_1$ is likely true). According to Relative Realism, then, the comparative ranking of competing theories provides sufficient grounds for believing that the theory that is judged to be superior to its competitors is the theory that is *closer* to the truth *relative* to its competitors. The comparative evaluation of competing theories, however, provides no reasons to believe that the theory that is judged to be superior to its competitors is close or closest to the truth, absolutely speaking. That is to say, we can never know how close to the truth our scientific theories are, absolutely speaking, because all we do in science is compare competing theories in terms of their empirical success.

This observation leads to another argument in support of Relative Realism, namely, the Argument from the Relative Success of Theories. That is to say, if the predictive success of scientific theories is *relative*, that is, $T_1$ is predictively successful only insofar as it is more predictively successful than its competitors $T_2$, $T_3$, ..., $T_n$, and predictive success is a property that makes a theory that has it better than a theory that does not have it, or a theory that has more predictive success is better than a theory that has predictive success but to a lesser degree, then it follows that, the more predictively successful a theory is, the better it is relative to its competitors. This means that the relative nature of predictive success warrants comparative judgments about competing theories (that is, $T_1$ is better than its competitors, $T_2$, $T_3$, ..., $T_n$, in terms of predictive success or $T_1$ is more predictively successful than its competitors, $T_2$, $T_3$, ..., $T_n$), rather than absolute judgements about competing theories (that is, $T_1$ is predictively successful). In that sense, an argument for Relative Realism can be made from the relative nature of predictive success. This argument can be stated in canonical (or standard) form as follows:

> (P1) The predictive success of scientific theories is relative (that is, $T_1$ is more predictively successful than its competitors $T_2$, $T_3$, ..., $T_n$).
>
> (P2) If the predictive success of scientific theories is relative (that is, $T_1$ is more predictively successful than its competitors $T_2$, $T_3$, ..., $T_n$), then scientists are justified in believing comparative judgements (that is, $T_1$ is more predictively successful than its competitors $T_2$, $T_3$, ..., $T_n$), rather than absolute judgements (that is $T_1$ is predictively successful), about competing theories.
>
> Therefore,
>
> (C) Scientists are justified in believing comparative judgements (that is, $T_1$ is more predictively successful than its competitors $T_2$, $T_3$, ..., $T_n$), rather than absolute judgements (that is $T_1$ is predictively successful), about competing theories.

As stated, this Argument from the Relative Success of Theories is a deductive argument (in particular, a *modus ponens*: If $A$, then $B$, $A$; therefore, $B$) for Relative Realism. Since the premises of this argument, namely, (P1) and (P2), successfully provide logically conclusive support for the conclusion, namely, (C), this argument can be said to be valid. The next question, then, is whether the premises are in fact true. Is the Argument from the Relative Success of Theories a sound argument for Relative Realism?

The premise according to which the predictive success of scientific theories is *relative*, namely, (P1) of the Argument from the Relative Success of Theories, is a corollary of the premise that theory evaluation is *comparative*, namely, (P1) of the Argument from the Comparative Evaluation of Theories. That is to say, if one agrees that scientists rank competing theories comparatively when they evaluate them, then one must also agree that a theory is predictively successful only insofar as it is more predictively successful relative to its competitors. This is because scientists evaluate theories by testing those theories' predictions empirically. More specifically, when scientists try to decide which theory provides the best explanation for some natural phenomena they seek to explain, they ask themselves what predictions those theories make, that is, what experimental results would be observed if $T_1$, $T_2$, $T_3$, ..., or $T_n$ were true. Depending on those observational and/or experimental results, scientists then select the theory among all the competing theories whose predictions are borne out by the results of observational and/or experimental testing. If $T_2$, say, makes more predictions that are borne out by observations and/or experiments than competing theories $T_1$ and $T_3$ do, then scientists are justified in judging that $T_2$ is more predictively successful than $T_1$ and $T_3$. And if scientists are justified in judging that $T_2$ is more predictively successful than $T_1$ and $T_3$, they are also justified in judging that $T_2$ is better than $T_1$ and $T_3$.

Now scientific realists should agree with (P1) of the Argument from the Relative Success of Theories. As we have seen in Chap. 4 (see Sect. 4.1), scientific realists advance the Positive Argument for scientific realism according to which approximate truth provides the best explanation for the *absolute* predictive success of scientific theories. If a scientific theory, $T$, can be predictively successful absolutely, then it follows that $T$ can also be predictively successful relative to its competing theories. Moreover, if the premise according to which the predictive success of scientific theories is *relative*, namely, (P1) of the Argument from the Relative Success of Theories, is a corollary of the premise that theory evaluation is *comparative*, namely, (P1) of the Argument from the Comparative Evaluation of Theories, as mentioned above, and scientific realists accept the latter, as we have seen in Sect. 6.3, then they should accept the former, too. Accordingly, if the Argument from the Relative Success of Theories is valid, and its premises are true, or at least acceptable to scientific realists, then it can be said to be a sound argument for Relative Realism.

Like scientific realists, antirealists should also accept the key premise of the Argument from the Relative Success of Theories, namely, (P1). For instance, according to Darrell Rowbottom (2019, p. 3949), who favors an antirealist position along the lines of Instrumentalism (see Chap. 3, Sect. 3.2), scientific realists must face the following challenge:

> What licenses inferring *absolute* confirmation values from *relative* confirmation values? If the realist cannot answer satisfactorily, it is reasonable to deny realism (emphasis added).

For Rowbottom (2019, p. 3949), the confirmation of a theory by its successful predictions (that is, those predictions that are borne out by observational and/or experimental testing), is "*relative* to the conceived alternatives" (emphasis in original). In other words, a theory is confirmed by its successful predictions relative to the competing theories with which it is actually compared. So it seems that Rowbottom would agree that confirmation by successful predictions is relative, which is precisely what (P1) of the Argument from the Relative Success of Theories states. The relative realist, in turn, would agree with Rowbottom (2019, p. 3949) that "*relative* confirmation has no established connection to truth-likeness, even on the assumption that *absolute* confirmation (in some non-subjective sense) does indicate truth-likeness (or probable truth-likeness, or whatever surrogate one prefers)" (emphasis added). This is why the relative realist stops short of claiming that we have good reasons to believe that our best scientific theories are approximately true or truth-like in the sense of being close to the truth. That is to say, according to the relative realist, we have no good reasons to believe *absolute* judgments about scientific theories, such as $T$ is well-confirmed, $T$ is predictively successful, $T$ is approximately true (that is, close to the truth), or $T$ is likely true. Rather, precisely because confirmation by testing the predictions a theory makes is relative, it can only provide good reasons to believe comparative judgments, such as $T_1$ is better confirmed than its competitors $T_2, T_3, …, T_n$, $T_1$ is more predictively successful than its competitors $T_2, T_3, …, T_n$, $T_1$ is closer to the truth than its competitors $T_2, T_3, …, T_n$; in other words, $T_1$ is comparatively true. So, when Rowbottom asks, "What licenses inferring *absolute* confirmation values from *relative* confirmation values?" the relative realist answers: Nothing licenses inferring *absolute* judgments about scientific theories from the *relative* nature of confirmation by predictive success. We must rest content with comparative judgments about our scientific theories. We must come to grips with the relative nature of scientific confirmation. The empirical confirmation and predictive success of scientific theories is inherently relative, that is, relative to the other theories with which those theories are compared.

Like Rowbottom (2019), it appears that Brad Wray (2010, p. 369) also agrees that "Success [...] is a *relative* notion" (emphasis added). That is to say, the "predictive success of our best theories is a *relative* success" (Wray 2010, p. 365). From this, it is safe to assume that Wray would also agree that a scientific theory is predictively successful relative to its competitors. In other words, we can say that a theory $T_1$ is more predictively successful than competitors $T_2, T_3, …, T_n$ precisely because the "predictive success of our best theories is a *relative* success" (Wray 2010, p. 365). As we have seen in Sect. 6.3, if theory evaluation, which is comparative in nature, allows us to judge that a theory $T_1$ is more predictively successful than competitors $T_2, T_3, …, T_n$, then it also allows us to judge that $T_1$ is *closer* to the truth than competitors $T_2, T_3, …, T_n$. This is because "a theory is, all else being equal, *better* if it can explain or predict facts from different domains" (Laudan 2004, p. 17) and "the superior theory is more likely true than the competitor theories with which it is compared" (Wray 2018, p. 45). It is evident, then, that antirealists like Laudan, Rowbottom, and Wray would accept the premises of the Argument from the Relative Success of Theories. Accordingly, if the Argument from the Relative Success of

Theories is valid, and its premises are true, or at least acceptable to antirealists, then it can be said to be a sound argument for Relative Realism.

Even if the Argument from the Comparative Evaluation of Theories (see Sect. 6.3) and the Argument from the Relative Success of Theories can be said to be sound arguments for Relative Realism, since they are valid arguments with premises that would be acceptable to both scientific realists and antirealists, it does not mean that the scientific realism/antirealism debate in contemporary philosophy of science is now settled. Far from it! As philosophers know all too well, there can be valid, and even sound, arguments supporting opposing philosophical positions. Nevertheless, I do think that these arguments for Relative Realism have several advantages over the arguments for scientific realism discussed in Chap. 4. The first advantage that the arguments for Relative Realism have over the arguments for scientific realism discussed in Chap. 4 is that they do not rely on Inference to the Best Explanation (IBE). As such, the arguments for Relative Realism are not open to the sort of criticisms typically leveled against arguments that make use of IBE. As we have seen in Chap. 4 (see Sect. 4.1), the key argument for scientific realism in contemporary philosophy of science is the Positive Argument for scientific realism, which is more commonly known as the "no miracles" argument. This argument is an instance of IBE. As we have seen, IBE is a non-deductive form of argumentation that proceeds from a phenomenon that requires an explanation to the conclusion that the best explanation for that phenomenon is probably true. The general form of IBE can be stated as follows:

1. Phenomenon $P$.
2. The best explanation for $P$ is $E$.
3. No other explanation explains $P$ as well as $E$ does.
4. Therefore, (probably) $E$.[7]

Notice, in particular, the third premise of an IBE, which presupposes a comparative evaluation of explanations, hypotheses, or theories. That is to say, in IBE, we are justified in concluding that the best explanation, hypothesis, or theory is the one we should accept only after we have compared it to competing explanations, hypotheses, or theories of the same phenomenon and determined that it is better than those alternative explanations, hypotheses, or theories. In that respect, the comparative nature of theory evaluation is built into IBE. Since scientific realists think that IBE is "ubiquitous in scientific practice" (Chakravartty 2017), and that IBE is "the inference that makes science" (McMullin 1992), they would have to agree that comparative theory evaluation is ubiquitous in science as well.[8] In other words, if IBE is an essential part of science (particularly, scientific reasoning), as scientific realists tend to believe, and the comparative nature of theory evaluation is an indispensable step in IBE, then it follows that the comparative nature of theory evaluation is an essential

---

[7] For more on the structure of IBE, see Psillos (2007).

[8] According to Igor Douven (2017), "philosophers of science have argued that abduction is a cornerstone of scientific methodology."

part of science. In that respect, science is inherently *comparative* and its results are intrinsically *relative*, just as the relative realist argues.

In contrast to scientific realists, antirealists tend to agree that "evaluations of competing theories are comparative in nature" (Wray 2018, p. 4), but they tend to reject IBE as an illegitimate form of inference (see Chap. 4, Sect. 4.1). The reason for this rejection of IBE has to do with an observation we have made before about the logical space of possible theories (see Sect. 6.3). Given the logical space of possible theories, we could be evaluating theories that are very far off from the truth. To use Bas van Fraassen's (1980, p. 143) terminology, when scientists select the best theory from a set of competing theories, they may simply be selecting the best of a "bad lot" (see Chap. 5, Sect. 5.2), just like selecting the best apple from a basket of rotten apples. So antirealists are unlikely to be impressed with arguments for scientific realism that make use of IBE.[9]

This "bad lot" objection would not work against the arguments for Relative Realism for the following reasons. First, the Argument from the Comparative Evaluation of Theories (see Sect. 6.3) and the Argument from the Relative Success of Theories are deductive arguments for Relative Realism, and they do not involve any Inferences to the Best Explanation (IBE). In that respect, unlike IBEs, which antirealists tend to find unpersuasive, the arguments for Relative Realism are deductive arguments that could, at least *in principle*, persuade antirealists. Second, the relative realist accepts the antirealist point about unconceived alternative theories that inhabit the logical space of possible theories. The relative realist agrees with antirealists that we do not know the exact position of our scientific theories in the logical space of possible theories, which means that the scientific theories we are selecting from might be theories that are very far away from the truth. Given that the relative realist argues from premises that antirealists would accept, without making any inferences that antirealists tend to find illegitimate, the Argument from the Comparative Evaluation of Theories (see Sect. 6.3) and the Argument from the Relative Success of Theories for Relative Realism have an advantage over arguments for scientific realism that take the form of IBE, which is a form of inference that antirealists tend to find unpersuasive.

The second advantage that the arguments for Relative Realism have over the arguments for scientific realism discussed in Chap. 4 is that they are not arguments that proceed from premises that state what are taken to be historical facts about science. As such, the arguments for Relative Realism are not open to the sort of criticisms typically leveled against arguments that proceed from premises that state what are taken to be historical facts about science. As we have seen in Chap. 4 (see Sect. 4.5), the Argument from the Exponential Growth of Science relies on historical facts about science (specifically, what Ludwig Fahrbach calls "the exponential growth of science"). However, such historical evidence is indeterminate between

---

[9] As we have seen in Chap. 5, Sect. 5.2, even though antirealists (specifically, constructive empiricists) reject IBE as an illegitimate form of inference in science, they sometimes fall back on IBE when giving a "positive argument" for their own antirealist position (specifically, Constructive Empiricism). See Mizrahi (2018).

realist and antirealist positions in the scientific realism/antirealism debate, which means that historical facts about science do not favor scientific realism over antirealism or antirealism over realism. That is to say, as philosophical theories about science, scientific realism and antirealism are observationally indistinguishable because they imply the same observational consequences about the history of science, as evidenced by the fact that scientific realists and antirealists often use historical evidence from the same historical record of science to support either scientific realism or antirealism, respectively. By the argument from the underdetermination of theories by evidence, then, scientific realism and antirealism are epistemically indistinguishable, given that they are equally well supported by the historical evidence. Consequently, there are no positive *historical* reasons to believe in scientific realism rather than antirealism. By contrast, the arguments for Relative Realism make no appeals to the historical record of science, which is indeterminate between realist and antirealist positions in the scientific realism/antirealism debate, and thus they have an advantage over arguments for scientific realism that do appeal to historical evidence.

Finally, the arguments for Relative Realism are not arguments from selected case studies cherry-picked from the history of science. As such, the arguments for Relative Realism are not open to the sort of criticisms typically leveled against arguments from selected case studies cherry-picked from the history of science. As we have seen in Chap. 2 (see Sect. 2.2), scientific realists and antirealists use case studies from the history of science to argue for their respective positions. Sometimes they even use the same case study to argue for scientific realism or for antirealism. Such is the case with the phlogiston case study. Antirealists have used it to argue against scientific realism (of some variety or another), whereas scientific realists have used it to argue for scientific realism (of some variety or another). Selected case studies cherry picked from the history of science, however, cannot provide a strong basis for inductive inferences about science as a whole. That is to say, samples that are too small or non-randomly selected (that is, cherry-picked) are not representative of science in general, and thus cannot provide strong inductive support for any conclusions about scientific realism or antirealism (see Chap. 5, Sect. 5.1). Indeed, as we have seen in Chap. 2 (see Sect. 2.2), drawing general conclusions or making predictions from small, non-random, and unrepresentative samples is a mistake in reasoning called "hasty generalization." According to Patrick Hurley (2006, p. 131), "Hasty generalization is a fallacy that affects inductive reasoning. [...] The fallacy occurs when there is a reasonable likelihood that the sample is not representative of the group. Such a likelihood may arise if the sample is either too small or not randomly selected." By contrast, the arguments for Relative Realism are not arguments from case studies, and thus they have an advantage over arguments for scientific realism that appeal to case studies.

To sum up, unlike other arguments for scientific realism (see Chap. 4), the arguments for Relative Realism, namely, the Argument from the Comparative Evaluation of Theories (see Sect. 6.3) and the Argument from the Relative Success of Theories for Relative Realism, are not IBEs, inductive generalizations from samples, or historical arguments from case studies. As such, they are not open to the sort of criticisms that are typically leveled against arguments for and against scientific realism that make use of IBE, samples, or case studies from the history of science. Accordingly, the relative realist argues from premises that both scientific realists and antirealists could accept (or at least can be reasonably expected to accept). In that respect, the arguments for Relative Realism have several advantages over other arguments for and against scientific realism.

## 6.5   Comparing Comparative Realism and Relative Realism

As we have seen in Sect. 6.1, the relative realist recommends reserving the term 'true' for theoretical claims, which are single or individual propositions that can be categorically true (or false), and using the term 'comparatively true' to talk about scientific theories, since 'comparative truth' expresses a comparative relation between competing theories. The notion of comparative truth is the core notion of Relative Realism, which is the view that, of a set of competing scientific theories, the more empirically successful theory is comparatively true, that is, closer to the truth relative to its competitors in the set. The Argument from the Comparative Evaluation of Theories (see Sect. 6.3) and the Argument from the Relative Success of Theories (see Sect. 6.4) are the key arguments for Relative Realism. Now, as mentioned in Sect. 6.2, Theo Kuipers (2019) has defended a view, which he calls "Comparative Realism." Given the apparent parallels between Comparative Realism and Relative Realism, it would be useful to compare the two.

Like Relative Realism, Comparative Realism recommends taking "the point of view of relations" and thinking about scientific theories "in terms of comparative notions, such as 'closer to the truth' and 'more successful than'," rather than absolute notions, such as 'true' and 'successful' (Kuipers 2019, p. 301). In that respect, according to both Relative Realism and Comparative Realism, we are only justified in making comparative judgements about the relative truth (or, as I call it, "comparative truth") of competing theories, such as $T_1$ is *closer* to the truth than competitors $T_2, T_3, ..., T_n$, not absolute judgements, such as $T_1$ is true. By way of illustration, Kuipers (2019, p. 306) gives the following example: Fresnel's wave theory of light is empirically *more successful than* its predecessor, namely, Huygens' wave theory of light, and its competitor, namely, Newton's particle emission theory of light. Unlike Relative Realism, however, the "main claim" of Comparative Realism, according to Kuipers (2019, p. 316), is that "truth approximation *provides the default explanation* and prediction of empirical and aesthetic progress" (emphasis added). For example, that Fresnel's wave theory of light is empirically *more successful than* its predecessor, namely, Huygens' wave theory of light, and its competitor, namely, Newton's particle emission theory of light, is supposed to be explained by the hypothesis that Fresnel's theory is *closer to the truth than* Huygens' theory and Newton's theory (Kuipers 2019, p. 306). This suggests that the argument for Comparative Realism is an Inference to the Best Explanation (IBE) of sorts according to which Comparative Realism is the default explanation of the empirical progress of science, an Inference to the Default Explanation (IDE), if you will. As Kuipers (2019, p. 301) himself puts it:

> The core of *comparative realism* becomes that the (comparative) phenomenon that one theory persistently is empirically more successful than another provides a good reason for the claim that the first theory is closer to the (theoretical and referential) truth than the second; a good, analytical, reason being that this very claim amounts to the generic *default explanation* for that comparative empirical phenomenon (emphasis in original).

Again, this suggests that the argument for Comparative Realism is an Inference to the Best Explanation (IBE) of sorts according to which the "default explanation" of the comparative empirical success of some scientific theory over another is that the former is closer to the truth than the latter. For convenience, let us call this form of argumentation "Inference to the Default Explanation" (IDE), so as to distinguish it from IBE.[10]

Now, Kuipers provides two versions of this Inference to the Default Explanation (IDE) in support of Comparative Realism. He labels the first version the "CR-no-miracles argument." According to Kuipers (2019, p. 305), the first version of the CR-no-miracles argument goes like this:

- As a rule, (persistently) empirically more successful theories are closer to the truth than, and refer at least as well as their predecessors.
- If this were not the case, the regular occurrence of persistently empirically more successful theories, including some novel predictive success, would be miraculous.
- Occasionally, other, case-specific, explanations of persistent empirically more successfulness, novel or not, may be appropriate.

> This cautious version of the CR-no-miracles argument suggests that truth approximation provides a kind of *default explanation and prediction* of (persistent) empirical more successfulness, including some novel predictive success (emphasis in original).[11]

As I understand it, the argument that Kuipers is making in this passage can be stated in canonical (or standard) form as follows:

(P1) There are scientific theories that are empirically more successful than their predecessors.
(P2) The phenomenon described in (P1) is explained by the hypothesis of truth approximation, that is, that empirically more successful theories are closer to the truth than their predecessors.

---

[10] IDE is also distinct from what Alexander Bird (2007b, p. 425) calls "Inference to the Only Explanation" (IOE), which is an inference to "the truth of some hypothesis since it is the only possible hypothesis left unrefuted by the evidence. It is the form of inference advocated by Sherlock Holmes in his famous dictum 'Eliminate the impossible, and whatever remains, however improbable, must be the truth'."

[11] When philosophers of science talk about "novel predictions," they typically mean a prediction that was not known to be true (or was expected to be true or false) at the time the theory was constructed.

(P3) The hypothesis of truth approximation provides the *default* explanation for the phenomenon described in (P1).

Therefore,

(C) The hypothesis of truth approximation is true.

Like the original "no-miracles" argument for scientific realism (see Chap. 4, Sect. 4.1), this CR-no-miracles argument for Comparative Realism is an instance of Inference to the Best Explanation (IBE) of sorts, but with a slight modification. As we have seen in Chap. 4 (Sect. 4.1), "Inference to the Best Explanation" (IBE)[12] "is sometimes also known as 'abduction'--following the terminology of Charles Peirce" (Ladyman 2002, p. 209). However, some philosophers have argued that IBE and Peirce's abduction are different forms of inference (Douven 2017).[13] Be that as it may, for Peirce, abduction is a non-deductive form of inference. Likewise, IBE is typically construed as an ampliative, or non-deductive, form of argumentation that proceeds from a phenomenon that requires an explanation to the conclusion that the best explanation for that phenomenon is probably true.[14] As James Ladyman (2007, p. 341) describes it, "Inference to the best explanation (IBE) is a (putative) rule of inference according to which, where we have a range of competing hypotheses all of which are empirically adequate to the phenomena in some domain, we should infer the truth of the hypothesis which gives us the best explanation of those phenomena." The general form of IBE can be stated as follows:

1. Phenomenon $P$.
2. The best explanation for $P$ is $E$.
3. No other explanation explains $P$ as well as $E$ does.
4. Therefore, (probably) $E$.[15]

In the case of the CR-no-miracles argument for Comparative Realism, the phenomenon that demands an explanation is "the regular occurrence of persistently empirically more successful theories" (Kuipers 2019, p. 305). This phenomenon "of (persistent) empirical more successfulness" (Kuipers 2019, p. 305) is supposed to be explained by the hypothesis of "truth approximation," that is, by the hypothesis that the (persistently) empirically more successful theories are closer to the truth than their predecessors. Unlike in IBE, however, the proposed hypothesis is not said to be the *best* among several competing hypotheses or explanations of the phenomenon "of (persistent) empirical more successfulness" (Kuipers 2019, p. 305). Rather,

---

[12] The phrase was coined by Gilbert Harman (1965).

[13] Daniel Campos (2011, p. 419) argues against the "tendency in the philosophy of science literature to link abduction to the inference to the best explanation (IBE), and in particular, to claim that Peircean abduction is a conceptual predecessor to IBE."

[14] For example, Alan Baker (2010, pp. 37–38) defines IBE as "A method of reasoning, also known as *abduction*, in which the truth of an hypothesis is inferred on the grounds that it provides the best explanation of the relevant evidence. In general, inference to the best explanation (IBE) is an ampliative (i.e., non-deductive) method" (emphasis in original).

[15] For more on the structure of IBE, see Psillos (2007).

the proposed hypothesis, namely, the "truth approximation" hypothesis, is said to be the *default* explanation of the phenomenon "of (persistent) empirical more success-fulness" (Kuipers 2019, p. 305). For, if the "truth approximation" hypothesis were not true, Kuipers (2019, p. 305) argues, the phenomenon "of (persistent) empirical more successfulness" would be miraculous.

The CR-no-miracles argument for Comparative Realism, then, is an Inference to the Default Explanation (IDE), if you will. Like IBE, IDE is a non-deductive form of argumentation. That is to say, the premises of the CR-no-miracles argument for Comparative Realism are not meant to provide logically conclusive support for its conclusion. Instead, they are meant to provide strong inductive support for its conclusion. Given that the CR-no-miracles argument for Comparative Realism is not meant to be a deductive argument, the next question is not whether it is valid or invalid, but rather whether it is strong or weak.

Unlike IBE (see Chap. 4, Sect. 4.1), which requires asserting that the proposed hypothesis is the *best* explanation among several competing explanations before concluding that the proposed hypothesis is probably true, Kuipers' IDE seems to require only that the proposed hypothesis is deemed the "default explanation" based on the claim that the phenomenon to be explained would be miraculous if the "default explanation" were not true. However, if there are alternative hypotheses, which explain the phenomenon in question just as well as the proposed hypothesis does, then one of those alternative hypotheses could be the default explanation. In that case, the inference from the default explanation to the correct (or likely true) explanation would be unwarranted. To see why this is so, an analogy might be help-ful. The default web browser of the Windows 10 operating system is Microsoft Edge. But the mere fact that Microsoft Edge is the default web browser is not a good reason to believe that it is the best, or even a good, web browser. In fact, most Internet users prefer web browsers other than Microsoft Edge. Fortunately, one can change the settings for the default web browser in the Windows 10 operating system and designate better web browsers, such as Mozilla Firefox (which offers the best performance, according to tech experts) or Google Chrome (which is the web browser that most Internet users prefer to use), as the default web browser. By anal-ogy, even if the "truth approximation" hypothesis is the default explanation for the phenomenon "of (persistent) empirical more successfulness" (Kuipers 2019, p. 305), this fact alone is not a good reason to believe that it is a good (let alone the best) explanation for that phenomenon. This is precisely why, in IBE, we need to compare competing explanations for the phenomenon in question before we can justifiably conclude that the best explanation is probably the correct one (see Sect. 6.4). Since Kuiper's IDE does not require such a comparison between competing explanations before concluding that the default explanation is probably the correct one, it seems that the conclusion that the proposed hypothesis is probably the cor-rect one is unwarranted.[16] If this is correct, then the CR-no-miracles argument can-not be said to be a strong argument for Comparative Realism.

---

[16] According to Kuipers (2019, p. 316), empirical progress occurs "when it is concluded on the basis of 'sufficient' *comparative* testing that the new theory is persistently empirically more suc-cessful than the old one" (emphasis added). Accordingly, Kuipers acknowledges that empirical

Moreover, as we have seen in Chap. 4 (see Sect. 4.1), constructive empiricists, like Bas van Fraassen, argue that Inference to the Best Explanation (IBE) is not a truth conducive form of argumentation. According to Bas van Fraassen (1980), from the fact that an explanation is the best one we could come up with, it does not follow that this explanation is probably true. For we might be working with a bad lot of explanations and the likely true explanation simply did not occur to us. So it is reasonable to expect that constructive empiricists would not be persuaded by Inferences to the Default Explanation (IDE), either. That is to say, constructive empiricists would argue that we might be working with a bad lot of explanations and the appropriate or default explanation simply did not occur to us. For this reason, even if a proposed hypothesis is deemed to be the default explanation for some phenomenon, it does not follow that it is probably the correct explanation of that phenomenon. Kuipers (2019, p. 318) would likely reply by claiming that "If the ['truth-approximation'] hypothesis is in fact false, new experiments will break the empirical progress conclusion in the long run." In other words, if a theory is not closer to the truth than its predecessors, further observational and experimental testing will eventually show that it is not as close to the truth as some other rival theory. However, constructive empiricists would insist that there is no need for the "truth approximation" hypothesis in order to account for either empirical progress or a break in empirical progress. As we have seen in Chap. 3 (see Sect. 3.3), for constructive empiricists, a theory is empirically adequate just in case what the theory says about what is observable (by us) is true. Consequently, as long as all the predictions of a theory are borne out by observational and experimental testing, the theory is considered empirically adequate, by the constructive empiricist's lights. If further observational and experimental testing shows that some of the theory's predictions are not borne out, then the theory is not considered empirically adequate (or, in comparative terms, it is less empirically adequate than its rival theories). There is no need to invoke "truth approximation" to account for this, or so constructive empiricists would argue.

Even if they were to accept IDE as a strong form of non-deductive inference, constructive empiricists would probably not accept the premises of the CR-no-miracles argument for Comparative Realism. In particular, they would object to (P3). In support of (P3), Kuipers claims that the occurrence of persistently empirically more successful theories would be miraculous if the "truth approximation" hypothesis were not true. But is it really the case that the phenomenon "of (persistent) empirical more successfulness" (Kuipers 2019, p. 305) would be miraculous if the hypothesis of "truth approximation" were not true? Well, the phenomenon "of (persistent) empirical more successfulness" (Kuipers 2019, p. 305) would be miraculous, at least to us, only if we cannot explain it. And if the hypothesis of "truth approximation" were the *only* explanation for the phenomenon "of (persistent) empirical more successfulness" (Kuipers 2019, p. 305), then it would be miraculous

---

testing or theory evaluation in science is comparative. By contrast, his IDE arguments for Comparative Realism do not seem to require a similar sort of comparative evaluation of theories.

if the phenomenon occurred and yet the only explanation for it were not true. The question, then, is whether the hypothesis of "truth approximation" is in fact the *only* explanation for the phenomenon "of (persistent) empirical more successfulness" (Kuipers 2019, p. 305), such that if it were not true, the phenomenon would be a miracle.[17] Kuipers (2019, p. 305) himself grants that "Occasionally, other, case-specific, explanations of persistent empirically more successfulness, novel or not, may be appropriate." But this seems to suggest that there are alternative explanations, and thus that the hypothesis of "truth approximation" is not the only explanation for the phenomenon "of (persistent) empirical more successfulness" (Kuipers 2019, p. 305). If that is the case, then it does not follow that the "truth approximation" hypothesis provides the default explanation for the regular occurrence of persistently empirically more successful theories from the claim that this phenomenon would be miraculous if the "truth approximation" hypothesis were not true. In other words, if there are other explanations of persistent empirically more successfulness, which may be appropriate on a case by case basis, as Kuipers (2019, p. 305) himself admits, then why is it that the "truth approximation" hypothesis is the *default* explanation? Unlike IBE,[18] Kuipers' IDE does not seem to require a comparative evaluation of alternative explanations before concluding that the "default explanation" is probably the correct one. So, for all we know, one of those other explanations could be the default explanation, since, absent a comparative evaluation of alternative explanations, the premises of the CR-no-miracles argument for Comparative Realism do not rule out this possibility. The question, then, is whether there are alternative explanations for the phenomenon "of (persistent) empirical more successfulness" (Kuipers 2019, p. 305).

As we have seen in Chap. 4 (see Sect. 4.1), antirealists have offered alternative explanations for the empirical success of science.[19] For example, according to van Fraassen (1980, pp. 39–40), the empirical success of our best scientific theories can be explained by positing a selection process for scientific theories that is similar to natural selection. As van Fraassen (1980, p. 40) puts it:

> The success of current scientific theories is no miracle. It is not even surprising to the scientific (Darwinist) mind. For any scientific theory is born into a life of fierce competition, a jungle red in tooth and claw. Only the successful theories survive--the ones which *in fact* latched on to actual regularities in nature (emphasis in original).

---

[17] Cf. Putnam's (1975, p. 73) claim that scientific realism "is the *only* philosophy that doesn't make the success of science a *miracle*" (emphasis added) and what Alexander Bird (2007b) calls "Inference to the Only Explanation" (IOE). See Chap. 4, Sect. 4.1.

[18] Or what Alexander Bird (2007b, p. 425) calls "Inference to the Only Explanation" (IOE).

[19] Kyle Stanford (2000) offers another antirealist explanation of the success of science in terms of "predictive similarity." According to Stanford, the predictive success of an abandoned theory can be explained by pointing out how closely its predictions approximate those of the accepted theory. For example, we "explain the success of the (revised) Ptolemaic system of epicycles by pointing out how closely its predictions approximate those of the true Copernican hypothesis. Let us call this relationship the *predictive similarity* of the Ptolemaic system to the Copernican" (Stanford 2000, p. 273).

Against scientific realists, who seek to explain the empirical success of our best scientific theories by means of the notion of approximate truth (see Chap. 4, Sect. 4.1), antirealists like van Fraassen offer an alternative explanation. For them, the fact that our best scientific theories are empirically successful can be explained by postulating a selection process by which unsuccessful theories are eliminated. This selection process is supposed to be akin to natural selection by which unfit species become extinct (Wray 2007).

This selectionist explanation for the empirical success of science can be extended to cover the phenomenon "of (persistent) empirical more successfulness" (Kuipers 2019, p. 305) as well. This is because the products of natural selection are not optimizations. Rather, natural selection is a "satisficing" process, not an optimizing process (Simon 1979, p. 3). Accordingly, if there is "a process of selection [for scientific theories] comparable to the selection process operative in the biological world" (Wray 2007, p. 81), then it should be a satisficing process as well, which means that it is the kind of process that brings about, not optimal results, but rather satisfactory results or results that meet a certain threshold. This threshold requires that new products (for example, species or theories) simply be better than previous ones, even though those new products could still be very far from optimal. Accordingly, like natural selection, whose products are simply better than previous ones (otherwise, they would not survive), the products of "a process of selection [for scientific theories] comparable to the selection process operative in the biological world" (Wray 2007, p. 81) would also be better than previous ones (otherwise, they would not survive). This is how the selectionist explanation can explain why a current theory is more empirically successful than a predecessor theory, for, if it were not more successful than its predecessor, it would not survive "a life of fierce competition" in science (van Fraassen 1980, p. 40). If this is correct, then (P3) of the CR-no-miracles argument for Comparative Realism cannot be said to be true, since constructive empiricists would argue that there are alternative explanations for the phenomenon "of (persistent) empirical more successfulness" (Kuipers 2019, p. 305), which means that, even if the hypothesis of "truth approximation" were not true, this phenomenon would not be miraculous because there would still be an explanation for it, namely, the selectionist explanation. A phenomenon for which we do have a natural explanation cannot be said to be miraculous. For this reason, the CR-no-miracles argument, even if strong, cannot be said to be a cogent argument for Comparative Realism.

- The same criticisms can be made against Kuipers' second version of his Inference to the Default Explanation (IDE) in support of Comparative Realism. He labels the second version the "CR-no-miracles argument, progress version." According to Kuipers (2019, pp. 305–306), the second version of the CR-no-miracles argument (the progress version) goes like this:
- As a rule, (persistently) empirical progress is due to theoretical truth approximation and referential truth preservation, if not referential truth approximation.
- If this were not the case, the regular occurrence of (novel) empirical progress would be miraculous.

- Occasionally, other, case-specific, explanations of (novel) empirical progress may be appropriate.

And, hence, in this form the argument suggests that (theoretical) truth approximation provides a kind of *default explanation and prediction of (novel) empirical progress* (emphasis in original).

As I understand it, the argument that Kuipers is making in this passage can be stated in canonical (or standard) form as follows:

(P1) There is empirical progress in science, that is, when a scientific theory is persistently empirically more successful than its predecessor.

(P2) The phenomenon described in (P1) is explained by the hypothesis of theoretical truth approximation, that is, an empirically more progressive theory is closer to the truth than its predecessor.

(P3) The hypothesis of theoretical truth approximation provides the *default* explanation for the phenomenon described in (P1).

Therefore,

(C) The hypothesis of theoretical truth approximation is true.

Like the first version of Kuipers' CR-no-miracles argument, the second version is also an IDE, so the same problems that afflict the first version afflict the progress version, too. First, as we have seen, even if a proposed hypothesis provides the default explanation for some phenomenon, that fact alone is not a good reason to believe that it is a good (let alone the best) explanation for that phenomenon. Recall the analogy to web browsers. Even if Microsoft Edge is the default web browser provided by the Windows 10 operating system, this fact alone is not a good reason to believe that Microsoft Edge is a good (let alone the best) web browser for Internet users. Indeed, both Internet users and tech experts consider Microsoft Edge to be inferior to alternative web browsers, such as Mozilla Firefox and Google Chrome. Second, constructive empiricists are unlikely to accept any argument that makes use of IDE for pretty much the same reasons they reject any argument that makes use of IBE. After all, the hypothesis that is deemed the "default explanation" of the phenomenon in question may simply be the "default explanation" of a "bad lot" of inappropriate explanations, none of which would be *the* default explanation for the phenomenon in question if appropriate alternative explanations had not been considered and evaluated comparatively. Finally, even if they did accept IDE (or IBE) as a legitimate form of argumentation, constructive empiricists would not accept (P3) of the second version of Kuipers' CR-no-miracles argument (the progress version) because they have an alternative explanation for empirical success, namely, the selectionist explanation.

As I see it, then, the main difference between Kuipers' Comparative Realism and Relative Realism lies in the ways in which these realist positions are argued for. Both versions of Kuipers' CR-no-miracles argument for Comparative Realism are instances of Inference to the Best Explanation (IBE) of sorts, or more precisely, Inference to the Default Explanation (IDE), whereas the arguments for Relative Realism are deductive arguments from the comparative nature of theory evaluation and the relative nature of predictive success. In other words, Kuipers' Comparative

Realism is wedded to IBE (or IDE) in a way that Relative Realism is not. Insofar as Kuipers' Comparative Realism is wedded to IBE (or IDE), it proceeds from premises and inferences that antirealists are unlikely to accept (or at least cannot be reasonably expected to accept). On the other hand, Relative Realism proceeds from premises and inferences that antirealists are likely to accept (or at least can be reasonably expected to accept). In that respect, unlike Comparative Realism, Relative Realism avoids the criticisms that are leveled against realist positions that rely on arguments like the Positive Argument (the so-called "no miracles" argument) for scientific realism (see Chap. 4, Sect. 4.1).

In fact, I would argue that we cannot justifiably infer scientific realism (of some variety or another) from explanatory and/or aesthetic considerations (see Mizrahi 2012). As we have seen in Chap. 4 (see Sect. 4.1), according to the Positive Argument for scientific realism (also known as the "no miracles" argument), scientific realism is probably true because it is the best explanation for the empirical success of our best scientific theories. The problem with this Inference to the Best Explanation (IBE) is that, as far as explanations go, scientific realism is not a very good explanation. First, there is at least one explanation, namely, the selectionist explanation, which seems to explain empirical success just as well as scientific realism does. If this is correct, then it is not the case that scientific realism provides the best explanation for the empirical success of our best scientific theories. Second, as we have seen in Chap. 4 (see Sect. 4.1), one of the good-making properties of explanations is unification. That is to say, good explanations are those that unify the phenomena to be explained, that is, they explain the most and leave the least unexplained things. As an explanation for the empirical success of science, however, scientific realism fails to unify the relevant phenomena. As Greg Frost-Arnold (2010, p. 51) argues:

> the empirical successes of general relativity, organic chemistry, and evolutionary biology provide evidence for the inductive generalization 'Mature Sciences Are Empirically Successful', but any substantive 'unifying work' is done at the level of this inductive generalization, which is the explanandum of scientific realism. No *further* unification is achieved by additionally claiming 'These mature sciences are (approximately) true' (emphasis in original).

In other words, the proposed hypothesis according to which our best scientific theories are approximately true (namely, scientific realism), does not do any unification work beyond what the inductive generalization from "Scientific theories $T_1, T_2,... T_n$ are empirically successful" to "Our best scientific theories are empirically successful" already does. For this reason, scientific realism cannot be said to be the *best* explanation for the empirical success of science because it is not even a *good* explanation. To be a good explanation, scientific realism must have some unification power, which it does not seem to have.

Third, as we have seen in Chap. 4 (see Sect. 4.1), another good-making property of explanations is testability. That is to say, good explanations are those that yield testable predictions, that is, consequences that can be tested by observation and experimentation. As an explanation for the empirical success of science, however, scientific realism fails to yield testable predictions. As Greg Frost-Arnold (2010, p. 47) argues, "adding the thesis of scientific realism to the conjunction of all our mature scientific theories (whichever those maybe) does not generate any novel predictions either, over and above the predictions already made by adding the generalization 'Mature Scientific Theories are Empirically Successful'." In other words, the proposed hypothesis according to which our best scientific theories are approximately true (namely, scientific realism), does not do any predictive work beyond what the inductive generalization from "Scientific theories $T_1$, $T_2$,... $T_n$ are empirically successful" to "Our best scientific theories are empirically successful" already does. For this reason, scientific realism cannot be said to be the *best* explanation for the empirical success of science because it is not even a *good* explanation. To be a good explanation, scientific realism must have some predictive power, which it does not seem to have.[20]

Finally, even if scientific realism did have some unification power and some predictive power, and thus would have been a good candidate for the best explanation of the empirical success of our best scientific theories, it would still be premature to conclude that it is the best explanation for empirical success. This is because scientific realism does not make independently testable predictions that, if true, would demonstrate its superiority to other explanations for empirical success. Consider an alternative explanation for empirical success we have already discussed, namely, the selectionist explanation. On their face, scientific realism and the selectionist explanations are plausible explanations of empirical success. How do we choose between these two explanations? Which one should we prefer? Are there good reasons to prefer one explanation over the other? The only reasonable way to choose between two plausible explanations of the same phenomenon is to independently test the predictions those explanations make. Here is an example that illustrates this point:

> A man is found in a cabin in a remote forest, with all the doors and windows securely locked from the inside, hanging dead from a noose. A suicide note lies on the table nearby. What would best explain this state of affairs?

Even though the man's death appears to be a clear case of suicide, there are other plausible explanations for his death. Consider the following alternative explanation:

> *Acting Accident*
> The man was rehearsing a drama about suicide, had locked the doors for privacy, and things had gone terribly wrong.

The Acting Accident hypothesis and the Suicide hypothesis are equally plausible explanations for the man's death because both are consistent with the available evidence. In other words, both hypotheses explain the facts before us, and the evidence does not leave the Suicide hypothesis as the only plausible explanation of the man's death. Of these two plausible explanations, then, which one counts as "the best"? We need a method of testing these two hypotheses whose results would yield a clear verdict that would favor one hypothesis over the other. Since both the Acting Accident hypothesis and the Suicide hypothesis seem to be plausible explanations

---

[20] See also Gerald Doppelt (2005, p. 1080), who asks, "What novel predictions do scientific realists make?"

for the man's death, it seems that the only way to choose between these competing hypotheses is to test their consequences or predictions independently. The Acting Accident hypothesis yields several predictions that can be tested independently of the hypothesis itself. For example, if the Acting Accident hypothesis were true, then the man would have owned a script, or would have been a member of a theatre group, or would have told his friends that he was involved with a new play. Whether or not the man was a member of a theatre group, for instance, can be checked independently of the Acting Accident hypothesis itself, for example, by calling local theatre groups and checking their members lists. Like the Acting Accident hypothesis, the Suicide hypothesis also yields several predictions that can be tested independently of the hypothesis itself. For example, if the Suicide hypothesis were true, then the man would have been under a lot of stress or would have been depressed prior to his death. Whether or not the man was suffering from stress or depression, for instance, can be checked independently of the Suicide hypothesis itself, for example, by checking his medical record, doctors' visits, any prescription drugs he may have been taking, interviewing his spouse and close friends, etc. Selecting the best explanation of the man's death, then, will depend on independently testing the predictions that each hypothesis yields. Suppose we find out that the man was recently fired from his job, had filed for bankruptcy, and got divorced. These findings would lend some support to the Suicide hypothesis. On the other hand, suppose we find out that the man was an actor, and that he had told his friends that he does not want to be disturbed this weekend because he needs to rehearse for an important role in a major play. These findings would lend some support to the Acting Accident hypothesis. At any rate, it would be the independent testing of predictions that would help us choose between these two plausible explanations for the same phenomenon.

To bring this point about independent testing of predictions home, consider an example from science. There are two plausible explanations for the expansion of the universe. The first is the big-bang model, according to which the universe underwent an explosion from a point of initial singularity. The second is the steady-state model, according to which the universe has always been expanding, which means that there was no explosion event, and its constant mean density is maintained by the continuous creation of matter. Both models explain the expansion of the universe equally well. Which one should we prefer? Are there good reasons to prefer one model over the other? Again, we need to test the consequences or predictions of these models independently. The big-bang model predicts the existence of relic radiation left over from the hot early phase of the big bang that permeates the entire universe. This radiation, which is now known as the Cosmic Microwave Background radiation (CMB), was discovered in 1965 by Arno Penzias and Robert Wilson. The CMB is an independently testable prediction of the big-bang model. That is to say, the existence of the CMB makes sense from the vantage point of the big-bang model, but not from the vantage point of the steady-state model, and that is why it counts as evidence in support of the former but not the latter.

Now, let us go back to our equally plausible explanations of empirical success, namely, scientific realism and the selectionist explanation, and ask the same questions cosmologists had asked about the big-bang model and the steady-state model. Which hypothesis is the best explanation for the empirical success of our best scientific theories? Do these two alternative explanations for empirical success yield independently testable predictions, which, if borne out by observational or experimental testing, would give us good reasons to prefer one over the other? Of course, both hypotheses "predict" the empirical success of our best scientific theories. So this "prediction" alone does not favor one explanation over the other. Other than this "prediction," however, it seems that scientific realism does not make any independently testable predictions such that, if borne out, would clearly favor it over the selectionist explanation for empirical success. And that is precisely the problem with scientific realism as an explanation for empirical success, that is, it "fails to yield independently testable predictions that alternative explanations for [empirical] success, [such as the selectionist explanation,] do not yield," which is why "there seems to be no good reason to prefer [scientific realism] over alternative explanations for success" (Mizrahi 2012, p. 137), such as the selectionist explanation.

In fact, if Brad Wray is right, then we have a good reason to prefer the selectionist explanation over scientific realism. That is to say, the selectionist explanation makes a testable prediction that scientific realism does not make. According to Wray, the selectionist explanation explains not only the empirical success of science, which scientific realism explains as well, but also the empirical failures of science. According to the selectionist explanation, there is a selection process of theories in science that is analogous to the selection process of species in nature. As Brad Wray (2007, p. 84) puts it:

> any theory that does not enable us to make accurate predictions is not apt to be around very long. No scientist will waste her career working with such a theory. As a result, *any theory that is still around is apt to be successful*. Consequently, when philosophers of science look at the world of science they should not be surprised to find only successful theories. The others have been eliminated or are on their way to being eliminated (emphasis added).

According to the selectionist explanation, then, scientific theories that fail to make accurate predictions are weeded out and replaced by those that succeed in doing so. Accordingly, this selectionist explanation would explain not only the empirical success of theories that survive, for if they were not empirically successful, this selection process would have eliminated them, but also the extinction of empirically unsuccessful theories, that is, those that were eliminated by the selection process. The latter is a phenomenon that scientific realism cannot explain. That is to say, just as there are scientific theories that are empirically successful, there are also scientific theories that are unsuccessful. Scientific realism can explain the empirical success of science, but it cannot explain the failures. Of course, that a scientific theory is approximately true would not explain why it is empirically unsuccessful. For, if it were approximately true, we would expect that theory to be empirically successful, not unsuccessful. As Wray (2018, p. 150) puts it, the selectionist explanation "provides us with the resources to explain the *failures* of science. The realist seems to have nothing to say here" (emphasis in original).

If Wray is right about this, then not only are there no good reasons to believe that scientific realism is the best explanation for empirical success, since scientific realism is not even a good explanation as it lacks unification power and predictive power, but there is actually a good reason to believe that the selectionist explanation is superior to scientific realism. This is because, unlike scientific realism, the selectionist explanation explains not only the empirical success of science but also the failures of science. Even if the selectionist explanation is not better than scientific realism as an explanation of both the success and the failures of science, it is still the case that scientific realism does not yield independently testable predictions that would clearly distinguish it as superior to alternative explanations for empirical success, such as the selectionist explanation. It is also the case that, as an explanation for empirical success, scientific realism lacks key properties that make explanations *good* explanations, such as unification power and predictive power. For these reasons, the prospects of defending scientific realism on the grounds that it provides the best explanation for the empirical success of our best scientific theories look grim.

## 6.6  Relative Realism as a Middle Ground Position

As we have seen in Sect. 6.4, Relative Realism should be an attractive position to antirealists, but it is still a robust enough realist position, I submit, for it preserves the realist's optimism about science's ability to get closer to the truth, which is characteristic of most versions of scientific realism (see Chap. 3). In that respect, I take Relative Realism to be a middle ground position that offers the best of both worlds: the realist's optimism about science's ability to get closer to the truth (that is, to make scientific progress), on the one hand, and the antirealist observations that theory evaluation is comparative and that predictive success is relative, on the other hand. However, unlike antirealists, who take (P1) of the Argument from the Comparative Evaluation of Theories (see Sect. 6.3) as a reason to withhold belief from judgements concerning the truth of theories, the relative realist takes it as a reason to withhold belief from *absolute* judgements, but not from *comparative* judgements, concerning the (comparative) truth of scientific theories, that is, judgments about how closer to the truth they are relative to their competitors.

As a further bonus to antirealists, Relative Realism seems to mesh nicely with one of their favorite explanations for the empirical success of science, namely, van Fraassen's selectionist explanation. As we have seen in Chap. 3 (see Sect. 3.1), both scientific realists and antirealists agree that science is successful. In the context of the scientific realism/antirealism debate in contemporary philosophy of science, the important kind of success is the predictive success exhibited by those scientific theories whose predictions are borne out by experimentation and observation. According to van Fraassen (1980, pp. 39–40), the predictive success of our best scientific theories can be explained by positing a selection process for scientific theories that is similar to natural selection. As van Fraassen (1980, p. 40) puts it:

> The success of current scientific theories is no miracle. It is not even surprising to the scientific (Darwinist) mind. For any scientific theory is born into a life of fierce competition, a jungle red in tooth and claw. Only the successful theories survive--the ones which *in fact* latched on to actual regularities in nature (emphasis in original).

Against scientific realists, who seek to explain the predictive success of our best scientific theories by means of the notion of approximate truth (see Chap. 4, Sect. 4.1), antirealists like van Fraassen offer an alternative explanation. For them, the fact that our best scientific theories are predictively successful can be explained by postulating a selection process by which unsuccessful theories are eliminated. This selection process is supposed to be akin to natural selection by which unfit species become extinct (Wray 2007).

As we have seen in Sect. 6.5, the products of natural selection are not optimizations. Rather, natural selection is a "satisficing" process, not an optimizing process (Simon 1979, p. 3). Accordingly, if there is "a process of selection [for scientific theories] comparable to the selection process operative in the biological world" (Wray 2007, p. 81), then it should be a satisficing process as well, which means that it is the kind of process that brings about, not optimal results, but rather satisfactory results or results that meet a certain threshold. This threshold requires that new products (for example, scientific theories) simply be better than previous ones (for example, scientific theories that are closer to the truth than their predecessors are), even though those new products could still be very far from optimal (for example, scientific theories that are very far from being absolutely true). It is evident, then, that antirealists should welcome this merging of Relative Realism with van Fraassen's selectionist explanation for the predictive success of our best scientific theories.

In that respect, it is important to note that Relative Realism is not meant to be an explanation for the predictive success of our best scientific theories. Nor is Relative Realism inferred as the best, only, or default explanation for such success. On Explanationist Realism, approximate truth is supposed to provide the best explanation for the predictive success of our best scientific theories (see Chap. 3, Sect. 3.1). In that respect, explanationist realists take the predictive success of a scientific theory to be a reliable indicator that the theory is approximately true (see Chap. 5, Sect. 5.1). As we have seen in Chap. 4 (see Sect. 4.1), however, there are various problems with the Positive Argument for scientific realism known as the "no miracles" argument. Indeed, as we have seen in Sect. 6.5, scientific realism lacks key properties that make explanations *good* explanations, such as unification power and predictive power. Moreover, scientific realism does not yield independently testable predictions that would clearly distinguish it from alternative explanations for empirical success, such as the selectionist explanation. This means that there are no good reasons to favor scientific realism over alternative explanations for success, such as the selectionist explanation. For both scientific realism and the selectionist explanation are equally plausible explanations for the predictive success of our best scientific theories. In fact, as we have seen in Sect. 6.5, there might actually be a good reason to believe that the selectionist explanation is superior to scientific realism because, unlike scientific realism, the selectionist explanation explains not only the empirical success of science but also the failures of science (Wray 2018, p. 150).

For these reasons, Seungbae Park's (2015, p. 26) complaint that Relative Realism "explains neither the success nor the failure of science" is rather misplaced. Park argues that Relative Realism fails to explain the success of science because comparative truth does not make success likely. As Park (2015, p. 23) puts it, "comparative truth is an inadequate explanatory property for success." As we have seen in Chap. 3 (see Sect. 3.1), on Explanationist Realism, approximate truth is supposed to provide the best explanation for the predictive success of our best scientific theories. In that respect, explanationist realists take the predictive success of a scientific theory to be a reliable indicator that the theory is approximately true (see Chap. 5, Sect. 5.1). On Relative Realism, by contrast, comparative truth is *not* supposed to provide the best explanation for the predictive success of our best scientific theories. Nor is the predictive success of our best scientific theories taken by the relative realist to be a reliable indicator that predictively successful theories are *comparatively true*. Unlike Explanationist Realism, Relative Realism is *not* inferred as the best explanation for the success of science. Instead, Relative Realism follows logically from the comparative nature of theory evaluation itself and from the relative nature of predictive success itself. Like Epistemic Structural Realism (ESR) (see Chap. 3, Sect. 3.5), Relative Realism falls under the epistemic dimension of scientific realism, since it is the view that our best scientific theories, in particular, those that are empirically successful, are not approximately true, but rather comparatively true relative to their competitors.

As far as the relative realist is concerned, how to explain the empirical success of our best scientific theories is still an open question. This is because the following are two distinct questions that, for the relative realist, can and should be kept apart:

(Q1) Do we have good reasons to believe what our best scientific theories say about unobservables?

(Q2) What is the best explanation of the empirical success of our best scientific theories?

Scientific realists, it seems, are interested in (Q2) insofar as they want to argue that their answer to (Q2), namely, approximate truth, allows them to answer (Q1) in the affirmative. That is to say, scientific realists want to argue for scientific realism on the grounds that it is the best explanation of empirical success. For instance, as we have seen in Chap. 3 (see Sect. 3.1), explanationist realists argue that we do have good reasons to believe what our best theories say about unobservables just in case those unobservables best explain the empirical success of the theories that posit them. And as we have seen in Chap. 4 (see Sect. 4.1), the Positive Argument for scientific realism is the argument that scientific realism is probably true because it provides the best explanation for the empirical success of our best scientific theories.

Given the problems with the Positive Argument for scientific realism (see Chap. 4, Sect. 4.1), and the fact that antirealists tend to be unmoved by Inferences to the Best Explanation (IBE), the relative realist avoids making arguments that rely on IBE. For this reason, (Q1) and (Q2) are two distinct questions, the answer to one does not bear on the answer to the other, as far as the relative realist is concerned. In fact, for the relative realist, the predictive success of our best scientific theories is not even an extraordinary fact that demands an explanation, let alone provide an argument for scientific realism as the best explanation for that fact. Given the logical space of possible theories, it is hardly surprising that, in exploring this space of theoretical possibilities, scientists sometimes discover theories that are empirically successful. For the space of theoretical possibilities is filled with all kinds of theories, with varying degrees of empirical success. To illustrate this point, an analogy might be helpful. According to the fine-tuning argument for the existence of God, our universe is "fine-tuned for life." To say that our universe is fine-tuned for life is to say that the values of certain cosmic parameters are such that, were they slightly different, life as we know it would not have existed in our universe. These cosmic parameters are the following (Manson 2009, p. 272):

| Parameter | Actual value |
|---|---|
| $M_p$ (mass of the proton) | 938.28 MeV |
| $M_n$ (mass of the neutron) | 939.57 MeV |
| c (the speed of light) | $2.99792458 \times 10^8$ m$^1$s$^{-1}$ |
| G (the Newtonian gravitational constant) | $6.6742 \times 10^{-11}$ m$^3$ kg$^{-1}$ s$^{-2}$ |

"Given this extremely improbable fine-tuning," the argument goes, "we should think it much more likely that God exists than we did before we learned about fine-tuning" (Manson 2009, p. 272). In other words, an extremely improbable occurrence, namely, the existence of life as we know it in our universe, demands an extraordinary explanation, namely, God, as the cause of that extremely improbable occurrence.

Now, a standard objection to the fine-tuning argument is to appeal to what is known as the "multiverse hypothesis" and argue that there is nothing extraordinary or extremely improbable about the fact that our universe is fine-tuned for life as we know it. For, according to the multiverse hypotheses, there are many (perhaps infinitely many) universes. Our universe is just one out of (perhaps infinitely) many universes (Greene 2011). If the multiverse hypothesis is true, then there is no need to posit the existence of God as an explanation for the fact that our universe is fine-tuned for life. For, if the multiverse hypothesis is true, then the probability of a fine-tuned universe is rather high, not extremely low, and thus there is no need to invoke a supernatural entity in order to explain the fact that our universe is fine-tuned for life (Bradley 2009). Similarly, if there is a logical space of infinitely many possible theories, then there is no need to posit approximate truth as an explanation for the fact that some of our theories are empirically successful. For, if there is a logical space of infinitely many possible theories, then the probability of finding an empirically successful theory as scientists explore this space of theoretical possibilities is rather high, not extremely low, and thus there is no need to invoke approximate truth in order to explain the fact that some of the theories scientists discover as they explore this space of theoretical possibilities are empirically successful. For in this space of theoretical possibilities, there are many theories that are empirically successful, many theories that are not empirically successful and many other theories with varying degrees of empirical success. There is nothing extraordinary or surprising about that fact, such that it demands an extraordinary explanation. In other words, the empirical success of our best scientific theories "is not even surprising to the scientific [(inflationist)] mind" (van Fraassen 1980, p. 40) of the relative realist.[21]

At any rate, the question about reasons to believe in theoretical knowledge, that is, (Q1), and the question about the best explanation for the empirical success of some scientific theories, that is, (Q2), are distinct questions, for the relative realist, that can and should be kept apart. As far as the relative realist is concerned, (Q2) remains an open question. There are at least two hypotheses that could explain this empirical success, and both are compatible with Relative Realism. The first hypothesis is the selectionist explanation. As mentioned above, the selection process for scientific theories is a "satisficing" process, so it produces new theories that are, not likely or approximately true, but rather *comparatively true*; that is, better (or *closer to the truth*) than their competitors. The second hypothesis, with which Relative Realism is compatible as well, is Kyle Stanford's explanation in terms of predictive similarity. According to Stanford (2000), the predictive success of an abandoned theory can be explained by pointing out how closely its predictions approximate those of the accepted theory. For example, according to Stanford (2000, p. 273), we "explain the success of the (revised) Ptolemaic system of epicycles by pointing out how closely its predictions approximate those of the true Copernican hypothesis. Let us call this relationship the *predictive similarity* of the Ptolemaic system to the Copernican" (emphasis in original). This antirealist explanation of the success of science is compatible with Relative Realism because it "does not appeal to a relationship between a theory and the world at all; instead it appeals to a relationship of predictive similarity *between two theories*" (Stanford 2000, p. 276; emphasis in original). As we have seen in Sect. 6.2, the relative realist would agree with Thomas Kuhn that the notion of a comparison between a theory and the world is "illusive in principle" (Kuhn 1962/1996, p. 206). The only warranted comparisons are those between competing theories. On the basis of such comparisons, we are entitled to rank competing theories in order: from more predictively successful to less predictively successful. Such comparative relations between competing theories, namely, *comparative truth* and *predictive similarity*, make "clear why it is no mystery or miracle that the successful theory enjoys the success that it does, without requiring that this theory itself be true" (Stanford 2000, p. 276), only that it is comparatively true relative to its competitors.

According to Relative Realism, then, theory evaluation provides sufficient warrant for comparative judgments, such as "$T_1$ is more predictively successful than competitors $T_2$, $T_3$, ... , $T_n$" and "$T_1$ is closer to the truth than competitors $T_2$, $T_3$, ... , $T_n$." This is why Relative Realism allows for scientific progress. Science makes progress by the introduction of new theories that are better or closer to the truth than

---

[21] According to Alan Guth's (1997) theory of cosmic inflation, or the "inflationary universe," the early universe has undergone a rapid exponential expansion the result of which was many "pocket" universes, or a "multiverse," where "the physical conditions vary greatly from pocket to pocket" (Steinhardt and Turok 2007, p. 227).

their earlier competitors. In that respect, Relative Realism retains what Wray (2018, p. 186) calls "a narrative of triumph," which is a realist narrative that antirealists "want to challenge." For this reason, there is something in Relative Realism for scientific realists, too. That is to say, Relative Realism does not challenge the realist's "narrative of triumph," as antirealists do. Instead, the relative realist embraces the realist's "narrative of triumph" about science and argues that science makes progress insofar as new theories are *comparatively* true *relative* to their earlier competitors. On Relative Realism, we may not have good reasons to believe that our scientific theories are *close* to the truth. But we do have good reasons to believe that our scientific theories are getting better, or *closer* to the truth, relative to their competitors, and hence that we are making scientific progress, comparatively speaking. Those reasons are: (a) the evaluation of competing theories is *comparative* in nature, and (b) the predictive success of our best scientific theories is *relative* to their competitors. If theory evaluation is comparative, and predictive success is relative, then all we are justified in believing is that our predictively successful theories are comparatively true (that is, *closer* to the truth than their competitors), but we are not justified in believing that our predictively successful theories are likely true, approximately true, or close to the truth, absolutely speaking.

For these reasons, too, Seungbae Park's (2015) complaint that Relative Realism is closer to skepticism than it appears to be is also misplaced. According to Park (2015, p. 19):

> Relative Realism is closer to skepticism about the claims of science than scientific antirealism is. Skepticism in this context is the view that none of what science says about the world is believable. The relative realist believes that a theory is *closer to the truth than its competitors*, though he does not believe that it is *close to the truth*. It is not clear, therefore, on what grounds the relative realist can believe that a theory is *approximately empirically adequate*, i.e., that most of its observational consequences are true. It seems that he can only believe that it is *comparatively empirically adequate*, i.e., that it is closer to empirical adequacy than its competitors. Clearly, then, the relative realist is more skeptical about what science says about the world than the scientific realist (emphasis added).

Park is right that, according to the relative realist, our more successful scientific theories are closer to the truth relative to their competitors, but not close to the truth, absolutely speaking. As we have seen in Sect. 6.3, from the comparative nature of theory evaluation, it follows that we are justified in believing comparative judgements (that is, $T_1$ is closer to the truth than competitors $T_2$, $T_3$, ..., $T_n$), rather than absolute judgements (that is, $T_1$ is likely true), about competing theories. Park is also right that, according to the relative realist, scientific theories are empirically successful, not absolutely, but only relative to their competitors. As we have seen in Sect. 6.4, from the relative nature of predictive success, it follows that we are justified in believing comparative judgements (that is, $T_1$ is more empirically successful than its competitors $T_2$, $T_3$, ..., $T_n$), rather than absolute judgements (that is $T_1$ is empirically successful), about competing theories. In that sense, the relative realist is indeed more circumspect than scientific realists in terms of what they are willing to believe about claims to theoretical knowledge in science. But that does not mean that the relative realist is a skeptic about theoretical knowledge in science. Unlike antirealists, who tend to deny the possibility of theoretical knowledge in science (see Chap. 2, Sect. 2.1), the relative realist holds that there is theoretical knowledge in science. It's just that theoretical knowledge in science is inherently *relative*. The relative realist would urge us to acknowledge the fact that theory choice in science is inherently comparative, that is, consisting of "choosing between competing theories" (Cordero 2000, p. 191). Likewise, the empirical confirmation and predictive success of scientific theories is inherently relative, that is, relative to the other theories with which those theories are compared, as well. Given that theory choice and theory confirmation in science are inherently relative, there is no escaping the conclusion that theoretical knowledge is inherently relative as well, or so the relative realist argues. However, the relativity of scientific theories does not imply skepticism about theoretical knowledge of unobservables. One can grant the possibility of theoretical knowledge in science, contrary to the skeptic, while, at the same time, accepting that theoretical knowledge in science is relative insofar as our best scientific theories are only comparatively true.[22]

In accordance with the realist's "narrative of triumph," when the relative realist looks at the historical record of science, he or she does not see a "graveyard of dead epistemic objects" (Chang 2011, p. 426), as many antirealists do. Instead, the relative realist sees some improvement insofar as some scientific theories are getting better relative to their earlier competitors. This is quite different from the ways in which scientific progress is typically understood by both scientific realists and antirealists, so it is worth explaining this dimension of Relative Realism in more detail. Typically, scientific progress is cached out in terms of some unit that is being accumulated or increased in quantity, such as truth, knowledge, or understanding. For example, according to Alexander Bird (2007a), scientific progress consists in the accumulation of scientific knowledge. As Bird (2008, p. 279) puts it, "An episode [in science] constitutes scientific progress precisely when it shows the accumulation of scientific knowledge." This view is known as the "epistemic account" of scientific progress according to which scientific progress consists in the accumulation of scientific knowledge. Other philosophers of science defend what is known as the "semantic account" of scientific progress according to which scientific progress consists in increasing approximation to truth. For example, according to Ilkka Niiniluoto (2014, p. 77), "scientific progress can be defined by increasing verisimilitude," that is, truthlikeness or closeness to the truth. (On approximate truth, see Chap. 2, Sect. 2.1). For Niiniluoto (2014, p. 75), "Historical case studies, which illustrate progress as increasing truthlikeness include shifts from Ptolemy to Snell's law of refraction," among others.[23] (On case studies as historical evidence, see Chap. 2, Sect. 2.2). Similarly, Bird (2008, p. 279) defines the semantic account of scientific progress as follows: "An episode [in science] constitutes scientific progress precisely when it either (a) shows the accumulation of true scientific belief, or (b)

---

[22] See Otávio Bueno's (2016) discussion of skepticism in the context of philosophy of science.

[23] Darrell Rowbottom has argued against the epistemic account and in defense of the semantic account of scientific progress. See his (2010) and (2015).

shows increasing approximation to true scientific belief." In contrast to both the epistemic account and the semantic account of scientific progress, Finnur Dellsén (2016) argues for an understanding-based account of scientific progress. According to the "noetic account" of scientific progress, science "makes cognitive progress precisely when it increases our understanding of some aspect of the world" (Dellsén 2018a, p. 451). As Dellsén (2018b, p. 7) explains, "Unlike Bird's epistemic account, [the noetic account] does not require that scientists have justification for, or even belief in, the explanations or predictions they propose."[24]

Accordingly, on all of the aforementioned accounts of scientific progress, the goal or aim of science is to accumulate more of, or increase the quantity of, some desired unit. On the epistemic account of scientific progress, the desired unit is knowledge and the goal or aim of science is to accumulate more, or to increase the quantity of, knowledge. On the semantic account of scientific progress, the desired unit is truth, truthlikeness, or verisimilitude, and the goal or aim of science is to accumulate more, or to increase the quantity of, truth, truthlikeness, or verisimilitude. On the noetic account of scientific progress, the desired unit is understanding and the goal or aim of science is to accumulate more, or to increase the quantity of, understanding.[25] Scientific progress is typically understood in this way, that is, as approaching a goal or aim by accumulating or increasing the quantity of some desired unit, probably because progress is thought to be "a *goal-relative* concept" (Niiniluoto 2019; emphasis in original). As Niiniluoto 2019) puts it, "Progress is a result-oriented concept, concerning the success of a product relative to some goal." This notion of scientific progress as goal-directed means that not all scientific research is successful in terms of attaining the goal of science. Accordingly, on the epistemic account of scientific progress, science is a knowledge-seeking enterprise and scientists sometimes succeed in making scientific progress when the results of their research add to the stockpile of scientific knowledge. On the semantic account of scientific progress, science is a truth-seeking enterprise and scientists sometimes succeed in making scientific progress when the results of their research add to the stockpile of scientific truth, truthlikeness, or verisimilitude. On the noetic account of scientific progress, science is an understanding-seeking enterprise and scientists sometimes succeed in making scientific progress when the results of their research add to the stockpile of scientific understanding.

---

[24] Seungbae Park has argued against the noetic account and in defense of the epistemic account of scientific progress. See his (2017b) and (2020).

[25] Another account of scientific progress that should be mentioned here is the "functional-internalist" account of scientific progress according to which "An episode [in science] shows scientific progress precisely when it achieves a specific goal of science, where that goal is such that its achievement can be determined by scientists at that time (e.g. solving scientific puzzles)" (Bird 2008, p. 279). Thomas Kuhn's account of scientific progress is a functional-internalist account of scientific progress, where "the solved problem is the basic unit of scientific progress" and "the aim of science is to maximize the scope of solved empirical problems" (Laudan 1977, p. 66). However, the current debate over the nature of scientific progress in contemporary philosophy of science has been focused mostly on the semantic, epistemic, and noetic accounts.

Now, if one takes the selectionist explanation for the empirical success of science seriously, as the relative realist does, then one cannot think of scientific progress in terms of the attainment of some fixed goal or aim. This is because, strictly speaking, an evolutionary process of natural selection is not a goal-directed process with a fixed, forward-looking goal or aim. As Stephen Jay Gould used to say, "evolution is nondirectional" (Prindle 2009, p. 72). Since the selectionist process of scientific theories is supposed to be analogous to natural selection, it cannot be a goal-directed process, either. Even though both the natural selection of species and the scientific selection of theories are not goal-directed processes, it does not necessarily mean that we cannot talk about the products of such selection processes as being *improvements* on their predecessors. As we have seen in Sect. 6.5, natural selection is a "satisficing" process, not an optimizing process (Simon 1979, p. 3). Accordingly, if there is "a process of selection [for scientific theories] comparable to the selection process operative in the biological world" (Wray 2007, p. 81), then it should be a *satisficing* process as well. A satisficing process is the kind of process that brings about, not optimal products, but rather satisfactory products or products that meet a certain threshold. This threshold requires that new products simply be *better than* previous ones, even though those new products could still be very far from optimal. In the case of natural selection, this satisficing process would bring about new species that would be better adapted to their environments than their predecessors, otherwise, they would not survive. In the case of the scientific selection of theories, this satisficing process would bring about new theories that would be more empirically successful than their predecessors, otherwise, they would not survive. For satisficing processes, whether natural or not, are "processes that select solutions to problems that are 'good enough' but can still have alternatives that are *even better*" (Fetzer 2010, p. 174; emphasis added).

Accordingly, a selection process analogous to natural selection does not preclude progress in terms of new products that are simply *better than* previous ones. In that sense, the relative realist can embrace the realist's "narrative of triumph" and argue that scientific progress occurs when scientific theories are getting better, that is, more empirically successful and comparatively true, relative to their earlier competitors. In evolutionary biology, "selection" can be defined as the "process allowing the proliferation of organisms that are relatively better adapted to external environmental conditions" (Ansdell and Hanson 2016, p. 179). Likewise, the relative realist would argue, scientific selection is the process allowing the proliferation of scientific theories that are relatively better adapted to observational and experimental testing than their earlier competitors were. In that respect, scientific selection, construed as a satisficing selection process that brings about theories that are more empirically successful and comparatively true relative to their earlier competitors, is consistent with the realist's "narrative of triumph" according to which science makes progress. Indeed, not only antirealists, like Bas van Fraassen, have found the evolutionary analogy useful for understanding science, but also scientific realists, like Karl Popper.[26] Popper (1972) sought to analyze the growth of scientific knowledge in terms of a selection process as well. On Popper's evolutionary model, scientific theories are subjected to rigorous testing by observation and experimentation. Those theories that do not survive the rigorous observational and experimental testing are discarded. Those theories that survive, do so merely for the time being. As in the natural world, the "life of fierce competition" in science (van Fraassen 1980, p. 40) is a constant reality for a scientific theory. Those theories that survive the rigorous testing merely "live to fight another day," so to speak.

Despite being compatible with the constructive empiricist's favorite explanation for the empirical success of science, namely, the selectionist explanation, or with Kyle Stanford's (2000) antirealist explanation in terms of predictive similarity, Relative Realism is still different from antirealist positions, such as Bas van Fraassen's Constructive Empiricism (see Chap. 3, Sect. 3.3) in at least two important respects. First, unlike constructive empiricists and other antirealists, the relative realist embraces the realist's "narrative of triumph" about science, as we have seen, and argues that science makes progress insofar as some new theories are *comparatively* true *relative* to their earlier competitors. Second, and related to the first point, scientific progress occurs when scientific theories improve upon earlier competitors and scientists are able to judge that such improvements have occurred by making judgments like the following:

(J1) $T_1$ is closer to the truth than its competitors $T_2$ and $T_3$.
(J2) $T_1$ is more successful than its competitors $T_2$ and $T_3$.

For the relative realist, comparative judgements like (J1) and (J2) can be justified on the basis of theory evaluation because, in evaluating theories, scientists rank the competitors comparatively. Constructive empiricists, on the other hand, would recommend agnosticism with respect to a comparative judgement like (J1). For constructive empiricists, (J1) sabotages its own possibility of vindication (van Fraassen 1989, p. 157), since (J1) is a judgement about theories that involve unobservables, which makes "vindication [...] *a priori* precluded" (van Fraassen 1983, p. 297). This is not to say that, for constructive empiricists, (J1) cannot be true. As van Fraassen (1989, p. 177) writes, "The truth of a judgement (for example, about unobservable entities) might not be decidable. Yet, a certain judgement may be true 'if only by accident'." Accordingly, although (J1) may be true "if only by accident," it is not decidable, and hence constructive empiricists would recommend suspending judgement about (J1). On the other hand, as far as the relative realist is concerned, theory evaluation, which is comparative by nature, can warrant belief in a comparative judgment like (J1). In other words, for the relative realist, the truth of a comparative judgment like (J1) is decidable by means of observational and experimental testing. As we have seen in Sect. 6.3, the Argument from the Comparative Evaluation of Theories is a valid argument for Relative Realism, and its premises should be

---

[26]Another antirealist who has found the analogy to evolution somewhat useful for understanding scientific development is Thomas Kuhn. James Marcum (2018) provides a useful discussion of the evolutionary elements in Kuhn's philosophy of science.

acceptable to both scientific realists and antirealists. This argument shows that theory evaluation can warrant belief in comparative judgments about the comparative truth of theories relative to their competitors, such as (J1). Moreover, as we have seen in Sect. 6.4, the Argument from the Relative Success of Theories is a valid argument for Relative Realism, and its premises should be acceptable to both scientific realists and antirealists. This argument shows that theory evaluation can warrant belief in comparative judgments about the comparative empirical success of theories relative to their competitors, such as (J2). Based on these two arguments for Relative Realism, then, we have adequate grounds for believing that our more empirically successful scientific theories are comparatively true relative to their earlier competitors. Science is making progress by means of theories that are *closer* to the truth relative to their competitors, for theory evaluation is comparative and predictive success is relative, it's just that we can never know how close to the truth our scientific theories are, absolutely speaking. There is no absolute frame of reference for the evaluation of competing scientific theories.

In that respect, it is important to note, as an anonymous reviewer encouraged me to do, that the relative realist's notion of comparative truth is not an epistemic notion of truth. An epistemic notion of truth is a notion of truth that "takes truth to be a form of rational acceptability" (Sankey 2008, p. 128). That is to say, on an epistemic notion of truth, a statement is true just in case it is rational to accept it as true. For constructive empiricists, no theoretical statement about unobservables is such that it is rational to accept it as true, which is why constructive empiricists would recommend agnosticism with respect to a comparative judgement like (J1), given that it is a claim to theoretical knowledge about unobservables. As we have seen in Chap. 3 (see Sect. 3.3), for constructive empiricists, to accept a scientific theory is to believe, not that it is (approximately) true, but rather that it is "empirically adequate," that is, that what the theory says about what is observable (by us) is true (van Fraassen 1980, p. 18).

On the other hand, a non-epistemic notion of truth is a notion of truth that takes truth to be some sort of "correspondence with reality" (Psillos 1999, p. 248). That is to say, on a non-epistemic notion of truth, a theoretical statement is true just in case what it says about reality corresponds with reality. As Howard Sankey (2008, p. 16) puts it, on the so-called correspondence theory of truth, "truth is a relation of correspondence that obtains in virtue of the world in fact being the way that it is said to be." As we have seen in Chap. 2 (see Sect. 2.1), this is the notion of truth commonly accepted by scientific realists. In that respect, the relative realist is in agreement with scientific realists. Where scientific realists and the relative realist part ways is in the ways in which they think about the status of scientific truths. As we have seen in Sect. 6.4, Darrell Rowbottom asks scientific realists to answer the following question: "What licenses inferring absolute confirmation values from relative confirmation values?" Scientific realists cannot answer this question, as Rowbottom rightly points out. This is because theory testing (or confirmation) is essentially comparative, and thus it can only license *relative* confirmation values (that is, $T_1$ is more experimentally and/or observationally confirmed than its competitors $T_2$ and $T_3$), not *absolute* confirmation values (that is, $T_1$ is experimentally and/or observationally confirmed). As Peter Lipton (1993, p. 89) puts it, theory "testing enables scientists to say which of the competing theories they have generated is likeliest to be correct, but does not itself reveal how likely the likeliest theory is." In other words, comparative theory testing entitles scientists to judge which theory of a set of competing theories is better confirmed and closer to the truth relative to its competitors in the set, but it does not entitle scientists to judge which theory is confirmed and close (or closest) to the truth, absolutely speaking.

It is important to emphasize, as we have seen in Sect. 6.2, that the relativized ranking of competing theories is different from the "big bugaboo" relativism that many philosophers of science think "must be defeated at all costs" (Nickles 2020) because it does not imply that "anything goes." For the relative realist, it is impossible for scientists to determine the *absolute truth* (or absolute confirmation value) of scientific theories, as the Kuhnian Argument from the Illusive Truth of Whole Theories purports to show (see Sect. 6.2). But it is possible for scientists to determine the *comparative truth* (or relative confirmation value) of theories relative to other competing theories. At an earlier stage of the scientific realism/antirealism debate in philosophy of science, Hilary Putnam argued for a view he called "Internal Realism." Even though that early stage of the debate is beyond the scope of this book, it may still be worth noting some of the differences between Relative Realism and Putnam's Internal Realism briefly. For Putnam's Internal Realism seems to imply that comparative evaluations of the sort the relative realist would argue are not only possible, but also justified by comparative theory evaluation, are not so according to Internal Realism.

First, Putnam's Internal Realism is notoriously difficult to pin down exactly, but it seems safe to say that Internal Realism amounts to the rejection of the metaphysical thesis (or stance or dimension) of scientific realism. As we have seen in Chap. 2 (see Sect. 2.1), "The *metaphysical stance* asserts that the world has a definite and mind-independent natural-kind structure" (Psillos 1999, p. xvii). Putnam's Internal Realism amounts to a rejection of this metaphysical thesis because it is meant to be a sharp contrast to metaphysical realism and because it urges us "to give up the picture of Nature as having its very own language which it is waiting for us to discover and use" (Putnam 1994, p. 302). By contrast, the relative realist does not reject the thesis that there is a mind-independent world out there. As we have seen, the relative realist's notion of comparative truth is a non-epistemic notion of truth. Presumably, some theories are more predictively successful than other competing theories, and some theories are closer to the truth than other competing theories, whether we know that or not. Second, as Yemima Ben-Menahem (2005, p. 5) explains, Putnam used the internal/external distinction to highlight the difference between "the human perspective and a super-perspective purporting to capture reality in itself." (Cf. footnote 6 on Perspectival Realism.) By contrast, comparative theory evaluation (that is, the relativized ranking of competing theories) does not require perspective-taking, according to the relative realist. That is to say, the relativized ranking of competing theories is not itself relativized to anything else, such as a perspective, a theory, or a framework. Rather, the relativized ranking of competing theories is relativized insofar as scientific theories can only be judged better or worse relative to their competitors. In other words, in science, theory evaluation (that is, the relativized ranking of competing theories) is inherently comparative, and thus it can only yield comparative knowledge.

Finally, an internal realist would probably reject the epistemic thesis of scientific realism as well. As we have seen in Chap. 2 (see Sect. 2.1), "The *epistemic stance* regards mature and predictively successful scientific theories as well-confirmed and approximately true of the world. So, the entities posited by them, or, at any rate, entities very similar to those posited, do inhabit the world" (Psillos 1999, p. xvii). Presumably, if "truth is an *idealization* of rational acceptability" (Putnam 1981, p. 55; emphasis in original), that is, truth is what is rational to accept "under ideal conditions" (Putnam 1981, p. 56), but the conditions of current science are not ideal, then it seems to follow that it would not be rational to accept that the "mature and predictively successful scientific theories [of current science are] well-confirmed and approximately true of the world" (Psillos 1999, p. xvii).[27] After all, current scientific theories are likely far from "the epistemically ideal theory produced at the ideal limit of scientific inquiry" (Sankey 2008, p. 128). Now, the relative realist would not endorse this epistemic thesis, either. Instead, the relative realist would argue that our more predictively successful theories are comparatively true, that is, closer to the truth relative to the competing theories they were evaluated against. Presumably, the internal realist would not endorse this epistemic thesis of Relative Realism because it requires taking "a super-perspective" (Ben-Menahem 2005, p. 5), which is impossible to do, or so the internal realist would argue. For the relative realist, however, replacing the problematic notion of approximate truth (see Chap. 2, Sect. 2.1) with the notion of comparative truth in formulating the epistemic thesis of Relative Realism also helps to make sense of what many scientific realists take for granted but have difficulty fitting into their realist positions, namely, the view that "science is fallible" (Rosenberg 2000, p. 112).[28] As Anjan Chakravartty (2017) puts it, the challenge scientific realists face is the following:

> realists are generally fallibilists, holding that realism is appropriate in connection with our best theories even though they likely cannot be proven with absolute certainty; *some of our best theories could conceivably turn out to be significantly mistaken*, but realists maintain that, granting this possibility, there are grounds for realism nonetheless (emphasis added).

But if there are grounds for scientific realism, that is, if we have good reasons to believe that our best scientific theories are *approximately true*, then how could those very theories still turn out to be *significantly mistaken*? Presumably, a scientific

---

[27] Admittedly, this is a matter of some debate in Putnam scholarship. This is partly because "Putnam's internal realism appears as *an alternative to a philosophical stance that he had stanchly defended*, i.e., metaphysical realism" (Silva 2008, p. 13; emphasis added). Also, recall that Putnam himself made a Positive Argument for scientific realism, the so-called "no miracles" argument (see Chap. 4, Sect. 4.1). Since the focus of this book is the scientific realism/antirealism debate in *contemporary* philosophy of science, however, we can gloss over these earlier stages of the debate.

[28] That science is fallible is a view shared by not only most scientific realists but also many practicing scientists. For example, according to the physicist, Allan Franklin (2002, p. 1), "we must remember that science is fallible."

theory that is significantly mistaken cannot also be approximately true. Rather than being approximately true, or close to the truth, a theory that is significantly mistaken is more likely approximately false. Accordingly, the challenge to scientific realists is to combine the epistemic thesis that our best scientific theories are approximately true with the view that those very theories could still turn out to be significantly mistaken (cf. Niiniluoto 2018).

The relative realist does not face this challenge. For the relative realist does not argue that our best scientific theories are approximately true. Instead, the relative realist argues that our best scientific theories, that is, those that are more empirically successful than their competitors, are comparatively true. As we have seen in Sect. 6.3, the relative realist would argue that, if scientists evaluate $T_1$, $T_2$, and $T_3$ by observational and experimental testing, they could reasonably make the comparative judgement that, say, $T_3$ is comparatively true relative to competitors $T_2$ and $T_1$ when $T_3$ outperforms $T_1$ and $T_2$ in such tests. However, a scientific theory can be closer to the truth relative to its competitors but still be quite far off from the truth, given the space of theoretical possibilities. Comparative theory evaluation cannot tell us which theory is *close* or *closest* to the truth, unless we have independent reasons to believe that the theories we are testing are those that are *close* or *closest* to the truth. Since we do not have independent reasons to believe that, we cannot reasonably claim that the theories we have tested are *close* or *closest* to the truth, although we can reasonably claim that one of them is *closer* to the truth than its competitors (that is, that a theory is comparatively true). In other words, theory evaluation can tell us which theory among competing theories is closer to the truth (for example, that $T_3$ is closer to the truth than $T_1$ and $T_2$). However, theory evaluation cannot tell us which theory among competing theories is close or closest to the truth. According to Relative Realism, then, since a scientific theory can be closer to the truth relative to its competitors, that is, comparatively true, but still be quite far off from the truth, a scientific theory can be comparatively true and still turn out to be significantly mistaken. There is no tension between the notion of comparative truth and the view that our best theories could turn out to be significantly mistaken as there is between the notion of approximate truth and that view. With its notion of comparative truth, then, Relative Realism fits nicely with the view, which is commonly accepted by many scientific realists as well as practicing scientists,[29] that science is fallible. Despite the fact that science is fallible, Greta Thunberg is right that we should "listen to the scientists" (see Chap. 1). Not because their best theories are true or approximately true. If the relative realist is right, there is no way for scientists to know that. Rather, we should listen to the scientists because their best theories are comparatively true, that is, they are closer to the truth than all the other theories with which they were compared. In that respect, "science is a work in progress, never completed" (Garrison and Ellis 2016, p. 7), as new scientific theories become more

---

[29] According to Sherri Roush (2010, p. 35), "There is no disagreement today that our science is fallible, that is, that we *could* be wrong, that our evidence is not strong enough to imply the truth of our theories. The issue is not whether we are fallible, but whether given that we are we can nevertheless have a right to confidence that our theories are true" (emphasis in original).

empirically successful and closer to the truth relative to their earlier competitors, but never absolutely so.


## 6.7   Summary

Relative Realism is a middle ground position between scientific realism and antirealism according to which, of a set of competing scientific theories, the more empirically successful theory is comparatively true, that is, the more empirically successful theory is closer to the truth relative to its competitors in the set. This brand of realism is not argued for by Inference to the Best Explanation (IBE) or by Inference to the Default Explanation (IDE), which is why it is not open to the criticisms leveled against the Positive Argument for scientific realism (also known as the "no miracles" argument) and those that can be leveled against Comparative Realism. Instead, the arguments for Relative Realism are deductive arguments from the comparative nature of theory evaluation and the relative nature of predictive success. The comparative nature of theory evaluation entails that we are justified in believing comparative judgements (that is, $T_1$ is closer to the truth than competitors $T_2, T_3, ..., T_n$), rather than absolute judgements (that is, $T_1$ is likely true), about competing theories (Sect. 6.3). Likewise, the relative nature of predictive success entails that we are justified in believing comparative judgements (that is, $T_1$ is more predictively successful than its competitors $T_2, T_3, ..., T_n$), rather than absolute judgements (that is $T_1$ is predictively successful), about competing theories (Sect. 6.4). According to Relative Realism, then, we have good reasons to believe that the more successful scientific theories are comparatively true, that is, they are closer to the truth relative to their competitors. As such, Relative Realism is a middle ground position between scientific realism and antirealism. On the one hand, the relative realist takes on board the antirealist's point that theory evaluation is comparative. On the other hand, the relative realist also embraces the realist's "narrative of triumph" about science and argues that science makes progress when new theories are *comparatively* true *relative* to their earlier competitors.


## Glossary

**Antirealism** An agnostic or skeptical attitude toward the theoretical posits (that is, unobservables) of scientific theories. Antirealism comes in different varieties, such as Constructive Empiricism (see Chap. 3, Sect. 3.3) and Instrumentalism (see Chap. 3, Sect. 3.2).

**Approximate truth** Closeness to the truth or truthlikeness. To say that a theory is approximately true is to say that it is close to the truth. According to some scientific realists, approximate truth is the aim of science. (See Chap. 2, Sect. 2.1).

**Case study** A particular, detailed description of a scientific activity, a scientific practice, or an episode from the history of science. (See Chap. 2, Sect. 2.2).

**Comparative truth** A relation between competing theories. To say that $T_1$ is comparatively true is to say that $T_1$ is closer to the truth than its competitors, $T_2$, $T_3$, ..., $T_n$. (See Sect. 6.1).

**Constructive Empiricism** The view that the aim of science is to construct empirically adequate theories. A theory is empirically adequate when what the theory says about what is observable (by us) is true. (See Chap. 3, Sect. 3.3).

**Empirical success** A scientific theory is said to be empirically successful just in case it is both explanatorily successful (that is, it explains natural phenomena that would otherwise be mysterious to us) and predictively successful (that is, it makes predictions that are borne out by observation and experimentation). (See Chap. 3, Sect. 3.1).

**The epistemic account of scientific progress** An account of scientific progress according to which progress in science consists in the accumulation of scientific knowledge. (See Sect. 6.6).

**The epistemic dimension (or stance) of scientific realism** The thesis that our best scientific theories, in particular, those that are empirically successful, are approximately true. (See Chap. 2, Sect. 2.1).

**Epistemic Structural Realism (ESR)** The view that the best scientific theories give us knowledge about the unobservable structure of the world. (See Chap. 3, Sect. 3.5).

**Explanatory success** A scientific theory is said to be explanatorily successful just in case it explains natural phenomena that would otherwise be mysterious to us. (See Chap. 3, Sect. 3.1).

**Hasty generalization** A fallacious inductive argument from a sample that is not representative of the general population that is the subject of the conclusion of the argument (because the sample is too small or cherry-picked rather than randomly selected). (See Chap. 2, Sect. 2.2).

**Inference to the Best Explanation (IBE)** An ampliative (or non-deductive) form of argumentation that proceeds from a phenomenon that requires an explanation to the conclusion that the best explanation for that phenomenon is probably true. (See Chap. 4, Sect. 4.1).

**The metaphysical dimension (or stance) of scientific realism** The thesis that there are things out there in the world for scientists to discover and that those things out there in the world are independent of the human minds that study them. (See Chap. 2, Sect. 2.1).

**Modus ponens** A form of argument with a conditional premise, a premise that asserts the antecedent of the conditional premise, and a conclusion that asserts the consequent of the conditional premise. That is, "if $A$, then $B$, $A$; therefore, $B$," where $A$ and $B$ stand for statements. *Modus ponens* is a valid form of inference, and so an argument in natural language that takes this logical form is valid. On the other hand, the following logical form is invalid: "if $A$, then $B$, $B$; therefore, $A$." It is known as the fallacy of affirming the consequent. (See Chap. 4, Sect. 4.1).

**Modus tollens** A form of argument with a conditional premise, a premise that denies the consequent of the conditional premise, and a conclusion that denies the antecedent of the conditional premise. That is, "if *A*, then *B*, not *B*; therefore, not *A*," where *A* and *B* stand for statements. *Modus tollens* is a valid form of inference, and so an argument in natural language that takes this logical form is valid. On the other hand, the following logical form is invalid: "if *A*, then *B*, not *A*; therefore, not *B*." It is known as the fallacy of denying the antecedent. (See Chap. 5, Sect. 5.1).

**The noetic account of scientific progress** An account of scientific progress according to which progress in science consists in increasing understanding. (See Sect. 6.6).

**Predictive success** A scientific theory is said to be predictively successful just in case it makes predictions that are borne out by observation and experimentation. (See Chap. 3, Sect. 3.1).

**Relative Realism** The view that, of a set of competing scientific theories, the more empirically successful theory is comparatively true, that is, closer to the truth relative to its competitors in the set. (See Sect. 6.1).

**Scientific realism** An epistemically positive attitude toward those aspects of scientific theories that are worthy of belief. Scientific realism comes in different varieties, such as Explanationist Realism (see Chap. 3, Sect. 3.1), Entity Realism (see Chap. 3, Sect. 3.4), Structural Realism (see Chap. 3, Sect. 3.5), and Relative Realism (see Sect. 6.1).

**Selectionist explanation** *for the success of science* An explanation for the empirical success of our best scientific theories according to which a selection process akin to natural selection by which fit species survive and unfit species go extinct explains the survival of successful theories and the extinction of unsuccessful theories. (See Chap. 4, Sect. 4.1).

**The semantic account of scientific progress** An account of scientific progress according to which progress in science consists in increasing approximation to truth or the accumulation of scientific truth. (See Sect. 6.6).

**Theoretical virtues** Properties of scientific theories, such as unification, testability, coherence, and simplicity, that make theories that have them good theories. Scientific realists tend to think of such properties as epistemic or truth conducive, whereas antirealists tend to think of them as merely pragmatic. (See Chap. 4, Sect. 4.1).

## References and Further Readings

Ansdell, M., & Hanson, C. A. (2016). Biogeography, microbial. In R. M. Kliman (Ed.), *Encyclopedia of evolutionary biology* (Vol. I, pp. 179–185). Amsterdam: Elsevier.

Ashton, N. A. (2020). Scientific perspectives, feminist standpoints, and non-silly relativism. In A. M. Crețu & M. Massimi (Eds.), *Knowledge from a human point of view* (pp. 71–85). Cham: Springer.

Baker, A. (2010). Inference to the best explanation. In F. Russo & J. Williamson (Eds.), *Key terms in logic* (pp. 37–38). London: Continuum.

Baxby, D. (1999). Edward Jenner's inquiry: A bicentenary analysis. *Vaccine, 17*(4), 301–307.

Ben-Menahem, Y. (2005). Introduction. In Y. Ben-Menahem (Ed.), *Hilary Putnam* (pp. 1–16). New York: Cambridge University Press.

Bird, A. (2007a). What is scientific progress? *Noûs, 41*(1), 64–89.

Bird, A. (2007b). Inference to the only explanation. *Philosophy and Phenomenological Research, 74*(2), 424–432.

Bird, A. (2008). Scientific progress as accumulation of knowledge: A reply to Rowbottom. *Studies in History and Philosophy of Science Part A, 39*(2), 279–281.

Bogen, J. (2020). Theory and observation in science. In E. N. Zalta (Ed.), *The Stanford encyclopedia of philosophy*, Summer 2020 Edition. https://plato.stanford.edu/archives/sum2020/entries/science-theory-observation/

Bradley, D. (2009). Multiple universes and observation selection effects. *American Philosophical Quarterly, 46*(1), 61–72.

Bueno, O. (2016). Epistemology and philosophy of science. In P. Humphreys (Ed.), *The Oxford handbook of philosophy of science* (pp. 233–251). New York: Oxford University Press.

Campos, D. G. (2011). On the distinction between Peirce's abduction and Lipton's inference to the best explanation. *Synthese, 180*(3), 419–442.

Chakravartty, A. (2017). Scientific realism. In E. N. Zalta (Ed.), *The Stanford encyclopedia of philosophy*, Summer 2017 Edition. https://plato.stanford.edu/archives/sum2017/entries/scientific-realism/

Chang, H. (2004). *Inventing temperature: Measurement and scientific progress.* New York: Oxford University Press.

Chang, H. (2011). The persistence of epistemic objects through scientific change. *Erkenntnis, 75*(3), 413–429.

Cordero, A. (2000). Realism, and the case of rival theories without observable differences. In E. Agazzi & M. Pauri (Eds.), *The reality of the unobservable: Observability, unobservability and their impact on the issue of scientific realism* (pp. 191–206). Dordrecht: Kluwer Academic Publishers.

Dellsén, F. (2016). Scientific progress: Knowledge versus understanding. *Studies in History and Philosophy of Science, 56*, 72–83.

Dellsén, F. (2018a). Scientific progress, understanding, and knowledge: Reply to Park. *Journal for General Philosophy of Science, 49*(3), 451–459.

Dellsén, F. (2018b). Scientific progress: Four accounts. *Philosophy Compass, 13*(11), e12525.

Doppelt, G. (2005). Empirical success or explanatory success: What does current scientific realism need to explain? *Philosophy of Science, 72*(5), 1076–1087.

Douven, I. (2017). Abduction. In E. N. Zalta (Ed.), *The Stanford encyclopedia of philosophy*, Summer 2017 Edition. https://plato.stanford.edu/archives/sum2017/entries/abduction/

Fenner, F., Henderson, D. A., Arita, I., Ježek, Z., & Ladnyi, I. D. (1988). *Smallpox and its eradication.* Geneva: World Health Organization. http://apps.who.int/iris/bitstream/10665/39485/1/9241561106.pdf

Fetzer, J. H. (2010). Is evolution an optimizing process? In E. Eells & J. H. Fetzer (Eds.), *The place of probability in science: In honor of Ellery Eells (1953–2006)* (pp. 163–179). Dordrecht: Springer.

Fine, A. (1986). Unnatural attitudes: Realist and instrumentalist attachments to science. *Mind, 95*(378), 149–179.

Franklin, A. (2002). *Selectivity and discord: Two problems of experiment.* Pittsburgh: University of Pittsburgh Press.

Frost-Arnold, G. (2010). The no-miracles argument for realism: Inference to an unacceptable explanation. *Philosophy of Science, 77*(1), 35–58.

Garrison, T., & Ellis, R. (2016). *Oceanography: An invitation to marine science* (9th ed.). Boston: Cengage Learning.

Giere, R. N. (2006). *Scientific perspectivism*. Chicago: The University of Chicago Press.

Greene, B. (2011). *The hidden reality: Parallel universes and the deep laws of the cosmos*. New York: First Vintage Books.

Guth, A. (1997). *The inflationary universe: The quest for a new theory of cosmic origins*. New York: Basic Books.

Harman, G. H. (1965). The inference to the best explanation. *The Philosophical Review, 74*(1), 88–95.

Hurley, P. J. (2006). *A concise introduction to logic* (9th ed.). Belmont: Wadsworth.

Jenner, E. (1800). *An inquiry into the causes and effects of the variolae vaccinae, a disease discovered in some of the western counties of England, particularly Gloucestershire, and known by the name of the Cow Pox* (2nd ed.). London: Sampson Low.

Kaufmann, S. H. E. (2008). Immunology's foundation: The 100-year anniversary of the Nobel prize to Paul Ehrlich and Elie Metchnikoff. *Nature Immunology, 9*, 705–712.

Kitcher, P. (1993). *The advancement of science: Science without legend, objectivity without illusions*. New York: Oxford University Press.

Kitcher, P. (2002). Scientific knowledge. In P. K. Moser (Ed.), *The Oxford handbook of epistemology* (pp. 385–407). New York: Oxford University Press.

Kuhn, T. S. (1962/1996). *The structure of scientific revolutions*. Chicago: The University of Chicago Press.

Kuipers, T. A. F. (2019). *Nomic truth approximation revisited*. Cham: Springer.

Kvanvig, J. (2003). *The value of knowledge and the pursuit of understanding*. New York: Cambridge University Press.

Ladyman, J. (2002). *Understanding philosophy of science*. London: Routledge.

Ladyman, J. (2007). Ontological, epistemological, and methodological positions. In T. Kuipers (Ed.), *General philosophy of science: Focal issues* (pp. 303–376). Amsterdam: Elsevier.

Laudan, L. (1977). *Progress and its problems: Towards a theory of scientific growth*. Berkeley: University of California Press.

Laudan, L. (2004). The epistemic, the cognitive, and the social. In P. Machamer & G. Wolters (Eds.), *Science, values, and objectivity* (pp. 14–23). Pittsburgh: University of Pittsburgh Press.

Leplin, J. (1997). *A novel defense of scientific realism*. New York: Oxford University Press.

Lim, D. (2003). *Microbiology* (3rd ed.). Dubuque: Kendall/Hunt Publishing Co.

Lipton, P. (1993). Is the best good enough? *Proceedings of the Aristotelian Society, 93*(1), 89–104.

Manson, N. A. (2009). The fine-tuning argument. *Philosophy Compass, 4*(1), 271–186.

Marcum, J. A. (2018). Revolution or evolution in science? A role for the incommensurability thesis? In M. Mizrahi (Ed.), *The Kuhnian image of science: Time for a decisive transformation?* (pp. 155–173). London: Rowman and Littlefield.

Massimi, M. (2004). Non-defensible middle ground for experimental realism: Why we are justified to believe in colored quarks. *Philosophy of Science, 71*(1), 36–60.

Massimi, M. (2018). Four kinds of perspectival truth. *Philosophy and Phenomenological Research, 96*(2), 342–359.

McMullin, E. (1992). *The inference that makes science*. Milwaukee: Marquette University Press.

Mizrahi, M. (2012). Why the ultimate argument for scientific realism ultimately fails. *Studies in History and Philosophy of Science Part A, 43*(1), 132–138.

Mizrahi, M. (2013). The argument from underconsideration and relative realism. *International Studies in the Philosophy of Science, 27*(4), 393–407.

Mizrahi, M. (2018). The "positive argument" for constructive empiricism and inference to the best explanation. *Journal for General Philosophy of Science, 49*(3), 1–6.

Musgrave, A. (2017). Strict empiricism versus explanation in science. In E. Agazzi (Ed.), *Varieties of scientific realism: Objectivity and truth in science* (pp. 71–94). Cham: Springer.

Nickles, T. (2020). Historicist theories of scientific rationality. In E. N. Zalta (Ed.), *The Stanford encyclopedia of philosophy*, Spring 2020 Edition. https://plato.stanford.edu/archives/spr2020/entries/rationality-historicist/

Niiniluoto, I. (2014). Scientific progress as increasing verisimilitude. *Studies in History and Philosophy of Science Part A, 46,* 73–77.

Niiniluoto, I. (2018). Scientific progress. In J. Saatsi (Ed.), *The Routledge handbook of scientific realism* (pp. 187–199). New York: Routledge.

Niiniluoto, I. (2019). Scientific progress. In E. N. Edward Zalta (Ed.), *The Stanford encyclopedia of philosophy,* Winter 2019 Edition. https://plato.stanford.edu/archives/win2019/entries/scientific-progress/

Park, S. (2015). Explanatory failures of relative realism. *Epistemologia, 38*(1), 16–28.

Park, S. (2017a). Critiques of minimal realism. *PRO, 92*(2017), 102–114.

Park, S. (2017b). Does scientific progress consist in increasing knowledge or understanding? *Journal for General Philosophy of Science, 48*(4), 569–579.

Park, S. (2020). Scientific understanding, fictional understanding, and scientific progress. *Journal for General Philosophy of Science, 51*(1), 173–184.

Popper, K. R. (1972). *Objective knowledge: An evolutionary approach.* Oxford: Clarendon Press.

Prindle, D. F. (2009). *Stephen Jay Gould and the politics of evolution.* New York: Prometheus Books.

Psillos, S. (1999). *Scientific realism: How science tracks truth.* London: Routledge.

Psillos, S. (2007). The fine structure of inference to the best explanation. *Philosophy and Phenomenological Research, 74*(2), 441–448.

Putnam, H. (1975). *Mathematics, matter and method.* New York: Cambridge University Press.

Putnam, H. (1981). *Reason, truth, and history.* New York: Cambridge University Press.

Putnam, H. (1994). In J. Conant (Ed.), *Words and life.* Cambridge, MA: Harvard University Press.

Rescher, N. (1990). *A useful inheritance: Evolutionary aspects of the theory of knowledge.* Savage: Rowman and Littlefield.

Rosenberg, A. (2000). *Philosophy of science: A contemporary introduction.* New York: Routledge.

Roush, S. (2010). Optimism about the pessimistic induction. In P. D. Magnus & J. Busch (Eds.), *New waves in philosophy of science* (pp. 29–58). Houndsmill: Palgrave Macmillan.

Rowbottom, D. P. (2010). What scientific progress is not: Against Bird's epistemic view. *International Studies in the Philosophy of Science, 24*(3), 241–255.

Rowbottom, D. P. (2015). Scientific progress without increasing verisimilitude: In response to Niiniluoto. *Studies in History and Philosophy of Science Part A, 51,* 100–104.

Rowbottom, D. P. (2019). Extending the argument from unconceived alternatives: Observations, models, predictions, explanations, methods, instruments, experiments, and values. *Synthese, 196*(10), 3947–3959.

Sankey, H. (2008). *Scientific realism and the rationality of science.* Hampshire: Ashgate.

Silva, C. C. (2008). Introduction: Putnam and the notion of "reality". In C. C. Silva, C. M. Vidal, & M. U. R. Monroy (Eds.), *Following Putnam's trail: on realism and other issues* (pp. 9–16). Amsterdam: Rodopi.

Simon, H. A. (1979). *Models of thought.* New Haven: Yale University Press.

Stanford, K. P. (2000). An antirealist explanation of the success of science. *Philosophy of Science, 67*(2), 266–284.

Steinhardt, P. J., & Turok, N. (2007). *Endless universe: Beyond the big bang.* New York: Doubleday.

Topper, D. R. (2013). *How Einstein created relativity out of physics and astronomy.* Dordrecht: Springer.

Tulodziecki, D. (2016). From zymes to germs: Discarding the realist/anti-realist framework. In R. Scholl & T. Sauer (Eds.), *The philosophy of historical case studies* (pp. 265–284). Basel: Springer.

van Fraassen, B. C. (1980). *The scientific image.* New York: Oxford University Press.

van Fraassen, B. C. (1983). Calibration: A frequency justification for personal probability. In R. S. Cohen & L. Laudan (Eds.), *Physics, philosophy and psychoanalysis* (pp. 295–319). Dordrecht: D. Reidel.

van Fraassen, B. C. (1989). *Laws and symmetry.* Oxford: Clarendon Press.

Wray, B. K. (2007). A selectionist explanation for the success and failures of science. *Erkenntnis, 67*(1), 81–89.

Wray, B. K. (2008). The argument from underconsideration as grounds for anti-realism: A defence. *International Studies in the Philosophy of Science, 22*(3), 317–326.
Wray, B. K. (2010). Selection and predictive success. *Erkenntnis, 72*(3), 365–377.
Wray, B. K. (2012). Epistemic privilege and the success of science. *Noûs, 46*(3), 375–385.
Wray, B. K. (2018). *Resisting scientific realism.* Cambridge: Cambridge University Press.

# Glossary

**Antirealism** An agnostic or skeptical attitude toward the theoretical posits (that is, unobservables) of scientific theories. Antirealism comes in different varieties, such as Constructive Empiricism (see Chap. 3, Sect. 3.3) and Instrumentalism (see Chap. 3, Sect. 3.2).

**Approximate truth** Closeness to the truth or truthlikeness. To say that a theory is approximately true is to say that it is close to the truth. According to some scientific realists, approximate truth is the aim of science. (See Chap. 2, Sect. 2.1.)

**Argument** A set of statements in which some (at least one statement called a premise) purport to provide logical support (either deductive or inductive) for another statement (namely, the conclusion). (See Chap. 1, Sect. 1.1.)

**Canonical form** A method of representing arguments where each premise is written on a separate, numbered line, followed by the conclusion (also known as "standard form"). (See Chap. 1, Sect. 1.1.)

**Case study** A particular, detailed description of a scientific activity, a scientific practice, or an episode from the history of science. (See Chap. 2, Sect. 2.2.)

**Cherry-picking** A sample from which an inductive inference is made is said to be cherry-picked when it is not randomly selected. (See Chap. 5, Sect. 5.1.)

**Circular definition** A definition is (viciously) circular when the term to be defined, or some variation thereof, is used in the definition itself. (See Chap. 4, Sect. 4.1.)

**Circular reasoning** An argument is (viciously) circular when its conclusion appears as one of the premises in the argument. Also known as begging the question. (See Chap. 4, Sect. 4.1.)

**Cogent argument** A strong argument with all true premises. (See Chap. 1, Sect. 1.1.)

**Comparative truth** A relation between competing theories. To say that $T_1$ is comparatively true is to say that $T_1$ is closer to the truth than its competitors, $T_2$, $T_3$, …, $T_n$. (See Chap. 6, Sect. 6.1.)

**Conclusion** The statement in an argument that the premises purport to support. (See Chap. 1, Sect. 1.1.)

© Springer Nature Switzerland AG 2020
M. Mizrahi, *The Relativity of Theory*, Synthese Library 431,
https://doi.org/10.1007/978-3-030-58047-6

**Constructive Empiricism** The view that the aim of science is to construct empirically adequate theories. A theory is empirically adequate when what the theory says about what is observable (by us) is true. (See Chap. 3, Sect. 3.3.)

**Deduction** A form of argumentation in which the premises purport to provide logically conclusive support for the conclusion. (See Chap. 1, Sect. 1.1.)

**Direct observation** Observation with the naked eye, without the use of scientific instruments, such as microscopes and telescopes, as opposed to instrument-aided observation. (See Chap. 3, Sect. 3.3.)

**Empirical adequacy** The aim of science, according to Constructive Empiricism. To say that a theory is empirically adequate is to say that what the theory says about what is observable (by us) is true. (See Chap. 3, Sect. 3.3.)

**Empirical success** A scientific theory is said to be empirically successful just in case it is both explanatorily successful (that is, it explains natural phenomena that would otherwise be mysterious to us) and predictively successful (that is, it makes predictions that are borne out by observation and experimentation). (See Chap. 3, Sect. 3.1.)

**Entity Realism** The view that the theoretical entities (that is, unobservables) posited by our best scientific theories are real. (See Chap. 3, Sect. 3.4.)

**The epistemic account of scientific progress** An account of scientific progress according to which progress in science consists in the accumulation of scientific knowledge. (See Chap. 6, Sect. 6.6.)

**The epistemic dimension (or stance) of scientific realism** The thesis that our best scientific theories, in particular, those that are empirically successful, are approximately true. (See Chap. 2, Sect. 2.1.)

**Epistemic Structural Realism (ESR)** The view that the best scientific theories give us knowledge about the unobservable structure of the world. (See Chap. 3, Sect. 3.5.)

**Explanationist Realism** The view that realist commitments are warranted with respect to the theoretical posits that are responsible for--or can best explain--the predictive success of our best scientific theories (also known as "Deployment Realism"). (See Chap. 3, Sect. 3.1.)

**Explanatory success** A scientific theory is said to be explanatorily successful just in case it explains natural phenomena that would otherwise be mysterious to us. (See Chap. 3, Sect. 3.1.)

**The exponential growth of science** The claim that scientific output grows at an exponential rate, with at least 95% of all scientific work having been done since 1915, and at least 80% of all scientific work having been done since 1950. (See Chap. 4, Sect. 4.5.)

**Fallacious argument** An argument whose premises fail to provide either conclusive or probable support for its conclusion (see also *invalid argument* and *weak argument*). (See Chap. 2, Sect. 2.2.)

**Hasty generalization** A fallacious inductive argument from a sample that is not representative of the general population that is the subject of the conclusion of the argument (because the sample is too small or cherry-picked rather than randomly selected). (See Chap. 2, Sect. 2.2.)

**The historical graveyard of science** The claim that, throughout the history of science, most scientific theories and theoretical posits have been abandoned, discarded, or replaced by new scientific theories and theoretical posits. (See Chap. 5, Sect. 5.1.)

**Induction** A form of argumentation in which the premises purport to provide probable support for the conclusion. (See Chap. 1, Sect. 1.1.)

**Inference to the Best Explanation (IBE)** An ampliative (or non-deductive) form of argumentation that proceeds from a phenomenon that requires an explanation to the conclusion that the best explanation for that phenomenon is probably true. (See Chap. 4, Sect. 4.1.)

**Instrument-aided observation** Observation by means of scientific instruments, such as microscopes and telescopes, as opposed to direct or naked-eye observation. (See Chap. 3, Sect. 3.3.)

**Instrumentalism** The view that scientific theories are instruments for attaining practical goals, such as predicting the occurrence of natural phenomena. (See Chap. 3, Sect. 3.2.)

**Invalid argument** A deductive argument in which the premises purport but fail to provide logically conclusive support for the conclusion. (See Chap. 1, Sect. 1.1.)

**The metaphysical dimension (or stance) of scientific realism** The thesis that there are things out there in the world for scientists to discover and that those things out there in the world are independent of the human minds that study them. (See Chap. 2, Sect. 2.1.)

**Modus ponens** A form of argument with a conditional premise, a premise that asserts the antecedent of the conditional premise, and a conclusion that asserts the consequent of the conditional premise. That is, "if $A$, then $B$, $A$; therefore, $B$," where $A$ and $B$ stand for statements. *Modus ponens* is a valid form of inference, and so an argument in natural language that takes this logical form is valid. On the other hand, the following logical form is invalid: "if $A$, then $B$, $B$; therefore, $A$." It is known as the fallacy of affirming the consequent. (See Chap. 4, Sect. 4.1.)

**Modus tollens** A form of argument with a conditional premise, a premise that denies the consequent of the conditional premise, and a conclusion that denies the antecedent of the conditional premise. That is, "if $A$, then $B$, not $B$; therefore, not $A$," where $A$ and $B$ stand for statements. *Modus tollens* is a valid form of inference, and so an argument in natural language that takes this logical form is valid. On the other hand, the following logical form is invalid: "if $A$, then $B$, not $A$; therefore, not $B$." It is known as the fallacy of denying the antecedent. (See Chap. 5, Sect. 5.1.)

**The noetic account of scientific progress** An account of scientific progress according to which progress in science consists in increasing understanding. (See Chap. 6, Sect. 6.6.)

**Ontic Structural Realism (OSR)** The view that everything that exists depends on the existence of structures and structures depend on nothing else for their existence. (See Chap. 3, Sect. 3.5.)

**Predictive success** A scientific theory is said to be predictively successful just in case it makes predictions that are borne out by observation and experimentation. (See Chap. 3, Sect. 3.1.)

**Premise** A statement in an argument that purports to support the conclusion of that argument. (See Chap. 1, Sect. 1.1.)

**The Problem of Unconceived Alternatives (PUA)** The claim that, throughout the history of science, scientists typically occupied an epistemic position in which they could conceive of only a few theories that were well-confirmed by the available evidence, while there were alternative theories that were as well-confirmed by the available evidence as those theories that were accepted by scientists. (See Chap. 5, Sect. 5.4.)

**Relative Realism** The view that, of a set of competing scientific theories, the more predictively successful theory is comparatively true, that is, closer to the truth relative to its competitors in the set. (See Chap. 6, Sect. 6.1.)

**Scientific realism** An epistemically positive attitude toward those aspects of scientific theories that are worthy of belief. Scientific realism comes in different varieties, such as Explanationist Realism (see Chap. 3, Sect. 3.1), Entity Realism (see Chap. 3, Sect. 3.4), Structural Realism (see Chap. 3, Sect. 3.5), and Relative Realism (see Chap. 6, Sect. 6.1).

**Selectionist explanation for the success of science** An explanation for the empirical success of our best scientific theories according to which a selection process akin to natural selection by which fit species survive and unfit species go extinct explains the survival of successful theories and the extinction of unsuccessful theories. (See Chap. 4, Sect. 4.1.)

**The semantic account of scientific progress** An account of scientific progress according to which progress in science consists in increasing approximation to truth or the accumulation of scientific truth. (See Chap. 6, Sect. 6.6.)

**The semantic dimension (or stance) of scientific realism** The thesis that scientific theories are to be taken literally, which means that they can be either true or false. (See Chap. 2, Sect. 2.1.)

**Sound argument** A valid argument with all true premises. (See Chap. 1, Sect. 1.1.)

**Strong argument** A non-deductive (or inductive) argument in which the premises successfully provide probable support for the conclusion. (See Chap. 1, Sect. 1.1.)

**Theoretical virtues** Properties of scientific theories, such as unification, testability, coherence, and simplicity, that make theories that have them good theories. Scientific realists tend to think of such properties as epistemic or truth conducive, whereas antirealists tend to think of them as merely pragmatic. (See Chap. 4, Sect. 4.1.)

**Underdetermination of theories by evidence** An antirealist argument according to which, if two theories are observationally indistinguishable, then they are epistemically indistinguishable, and thus there are no positive reasons to believe in one over the other. (See Chap. 4, Sect. 4.5.)

**Valid argument** A deductive argument in which the premises successfully provide logically conclusive support for the conclusion. (See Chap. 1, Sect. 1.1.)

**Weak argument** A non-deductive (or inductive) argument in which the premises purport but fail to provide probable support for the conclusion. (See Chap. 1, Sect. 1.1.)

# Author Index

**B**
Ben-Menahem, Y., 153, 154
Berenstain, N., 3–6
Bird, A., 53, 94, 131, 135, 148, 149
Bueno, O., vi, 44, 148

**C**
Chakravartty, A., vi, 2, 21, 23, 26, 81, 111,
    127, 154
Chang, H., 81, 84, 86, 100, 117, 148
Cordero, A., vi, 45, 123, 148

**D**
Dellsén, F., 149
Duhem, P., 36, 38

**E**
Ehrlich, P., 112
Einstein, A., 36, 37, 41, 42, 115

**F**
Fahrbach, L., 29, 52, 69, 70, 72, 73,
    82, 83, 128
Fine, A., 56, 110
French, S., vi, 36, 44, 45

**G**
Giere, R., 119
Godfrey-Smith, P., 96, 97

**H**
Hacking, I., 36, 42, 43, 52, 66
Hesse, M., 92, 117

**I**
Ivanova, M., 25, 38, 72

**J**
Jenner, E., 111, 112

**K**
Kitcher, P., 23, 30, 36, 61, 62, 111
Kuhn, T., 115–118, 120, 146, 149, 151
Kuipers, T.A.F., 119, 130–137

**L**
Ladyman, J., 5, 36, 45, 53, 54, 56,
    81, 88, 132
Laudan, L., 25, 56, 80–82, 86, 122, 123,
    126, 149
Leplin, J., 23, 111
Lipton, P., 80, 81, 100, 122, 123, 152
Lyons, T., vi, 37, 84, 85

**M**
Marcum, J.A., 151
Massimi, M., 68, 117, 119
Maxwell, G., 45, 52, 63–66
McMullin, E., 127

© Springer Nature Switzerland AG 2020
M. Mizrahi, *The Relativity of Theory*, Synthese Library 431,
https://doi.org/10.1007/978-3-030-58047-6

# Subject Index

## A

Approximate truth, 23, 24, 36, 56, 74, 85–87, 103, 104, 111–114, 125, 136, 143–145, 148, 154–156, 163
Argumentation, v, 3, 4, 7, 8, 10, 13–15, 25, 29, 31, 54, 67, 74, 88–90, 109, 127, 131–134, 137, 157, 164, 165
Astronomy, 9, 43, 86, 101, 102
Astrophysics, 43

## B

Bad lot, 54, 67, 89, 128, 134, 137
Begging the question, 56, 74, 163
Black holes, 36, 41–43

## C

Case study, 12, 26–32, 71, 72, 129
Circular definition, 57, 58
Circular reasoning, 56, 57
Climate change, 1, 13, 19, 20
Comparative Realism, 119, 130–138, 156
Comparative truth, 114–116, 118, 120, 124, 130, 144, 146, 152–155
Confirmation bias, 97
Constructive Empiricism, 36, 39–41, 46, 47, 55, 72, 73, 80, 87–90, 103, 104, 128, 151, 156, 157
Continental drift, 11, 24
Coronavirus, 12, 39, 97
Corroboration, 43, 66–68
Counterexample, 9, 80, 81, 86, 87, 103
COVID-19, 12, 22, 39, 97

## D

Deoxyribonucleic acid (DNA), 20, 43, 66, 87, 92, 116
Deployment Realism, 37, 47
Detection, 40, 66–68, 73

## E

Empirical adequacy, 36, 40, 46, 47, 61, 62, 103, 104, 147, 164
Empirical success, 24, 36, 47, 52–56, 58–60, 62, 73, 80, 124, 131, 135–146, 150–152, 157
Entity Realism, 25, 36, 42–44, 52, 66–68, 75, 158
Epistemic account of scientific progress, 149, 157, 164
Epistemic privilege, 92, 94
Event Horizon Telescope, 42, 93
Evolution, 19, 20, 150, 151
Explanationist Realism, 25, 36–38, 44, 47, 48, 72, 88, 90, 143, 144, 158, 164, 166
Exponential growth of science, 69–74, 83, 128

## G

Gallup poll, 1
General Relativity, 36, 37, 41, 42, 138
Germs, 12, 113
Global warming, 1, 2, 20
"Graveyard" Argument, 69, 81, 83–87, 92, 94, 100

© Springer Nature Switzerland AG 2020
M. Mizrahi, *The Relativity of Theory*, Synthese Library 431,
https://doi.org/10.1007/978-3-030-58047-6

Printed in the United States
by Baker & Taylor Publisher Services